Student Solutions Manual and Study Guide

for
Serway and Faughn's

College Physics

Seventh Edition
Volume 2

John R. Gordon
James Madison University

Charles Teague
Eastern Kentucky University

Raymond A. Serway
Emeritus, James Madison University

THOMSON

BROOKS/COLE

Australia • Canada • Mexico • Singapore • Spain • United Kingdom • United States

Printed in the United States of America
2 3 4 5 6 7 09 08 07 06

Printer: Darby Printing Company

0-534-99930-1

For more information about our products, contact us at:
Thomson Learning Academic Resource Center
1-800-423-0563

For permission to use material from this text or product, submit a request online at
http://www.thomsonrights.com.
Any additional questions about permissions can be submitted by email to **thomsonrights@thomson.com.**

Thomson Higher Education
10 Davis Drive
Belmont, CA 94002-3098
USA

Asia (including India)
Thomson Learning
5 Shenton Way
#01-01 UIC Building
Singapore 068808

Australia/New Zealand
Thomson Learning Australia
102 Dodds Street
Southbank, Victoria 3006
Australia

Canada
Thomson Nelson
1120 Birchmount Road
Toronto, Ontario M1K 5G4
Canada

UK/Europe/Middle East/Africa
Thomson Learning
High Holborn House
50–51 Bedford Road
London WC1R 4LR
United Kingdom

Latin America
Thomson Learning
Seneca, 53
Colonia Polanco
11560 Mexico
D.F. Mexico

Spain (including Portugal)
Thomson Paraninfo
Calle Magallanes, 25
28015 Madrid, Spain

Table of Contents

Preface

This *Student Solutions Manual and Study Guide* has been written to accompany the textbook, *College Physics, Seventh Edition*, by Raymond A. Serway and Jerry S. Faughn. The purpose of this ancillary is to provide students with a convenient review of the basic concepts and applications presented in the textbook, together with solutions to selected end-of-chapter problems. This is not an attempt to rewrite the textbook in a condensed fashion. Rather, emphasis is placed upon clarifying typical troublesome points, and providing further practice in methods of problem solving.

Every textbook chapter has a matching chapter in this book and each chapter is divided into several parts. Very often, reference is made to specific equations or figures in the textbook. Each feature of this Study Guide has been included to insure that it serves as a useful supplement to the textbook. Most chapters contain the following components:

- **Notes From Selected Chapter Sections:** This is a summary of important concepts, newly defined physical quantities, and rules governing their behavior.

- **Equations and Concepts:** This is a review of the chapter, with emphasis on highlighting important concepts and describing important equations and formalisms.

- **Suggestions, Skills, and Strategies:** This section offers hints and strategies for solving typical problems that the student will often encounter in the course. In some sections, suggestions are made concerning mathematical skills that are necessary in the analysis of problems.

- **Review Checklist:** This is a list of topics and techniques the student should master after reading the chapter and working the assigned problems.

- **Solutions to Selected End-of-Chapter Problems:** Solutions are given for selected odd-numbered problems that were chosen to illustrate important concepts in each textbook chapter.

- **Tables:** A list of selected Physical Constants is printed on the inside front cover; and a table of some Conversion Factors is provided on the inside back cover.

An important note concerning significant figures: The answers to most of the end-of-chapter problems are stated to three significant figures, though calculations are carried out with as many digits as possible. We sincerely hope that this *Student Solutions Manual and Study Guide* will be useful to you in reviewing the material presented in the text, and in improving your ability to solve problems and score well on exams. We welcome any comments or suggestions which could help improve the content of this study guide in future editions; and we wish you success in your study.

John R. Gordon
Harrisonburg, VA

Charles Teague
Richmond, KY

Raymond A. Serway
Leesburg, VA

Acknowledgments

We take this opportunity to thank everyone who contributed to this Seventh Edition of *Student Solutions Manual and Study Guide to Accompany College Physics*.

Special thanks for managing and directing this project go to Editorial Assistant Brandi Kirksey, Developmental Editor Ed Dodd, Acquisitions Editor Chris Hall, and Publisher for the Physical Sciences David Harris.

Our appreciation goes to our reviewer, Joseph A. Keane of St. Thomas Aquinas College. His careful reading of the manuscript, offering many useful suggestions, and checking the accuracy of the problem solutions contributed in an important way to the quality of the final product. Any errors remaining in the manual are the responsibility of the authors.

It is a pleasure to acknowledge the excellent work of Martin Arthur of Atelier 88 who prepared the final page layout and camera-ready copy. His technical skills and attention to detail added much to the appearance and usefulness of this volume.

Finally, we express our appreciation to our families for their inspiration, patience, and encouragement.

Suggestions for Study

We have seen a lot of successful physics students. The question, "How should I study this subject?" has no single answer, but we offer some suggestions that may be useful to you.

1. Work to understand the basic concepts and principles before attempting to solve assigned problems. Carefully read the textbook before attending your lecture on that material. Jot down points that are not clear to you, take careful notes in class, and ask questions. Reduce memorization to a minimum. Memorizing sections of a text or derivations does not necessarily mean you understand the material.

2. After reading a chapter, you should be able to define any new quantities that were introduced and discuss the first principles that were used to derive fundamental equations. A review is provided in each chapter of the Study Guide for this purpose, and the marginal notes in the textbook (or the index) will help you locate these topics. You should be able to correctly associate with each physical quantity the symbol used to represent that quantity (including vector notation, if appropriate) and the SI unit in which the quantity is specified. Furthermore, you should be able to express each important principle or equation in a concise and accurate prose statement. Perhaps the best test of your understanding of the material will be your ability to answer questions and solve problems in the text, or those given on exams.

3. Try to solve plenty of the problems at the end of the chapter. The worked examples in the text will serve as a basis for your study. This Study Guide contains detailed solutions to about twelve of the problems at the end of each chapter. You will be able to check the accuracy of your calculations for any odd-numbered problem, since the answers to these are given at the back of the text.

4. Besides what you might expect to learn about physics concepts, a very valuable skill you can take away from your physics course is the ability to solve complicated problems. The way physicists approach complex situations and break them down into manageable pieces is widely useful. Starting in Section 1.10, the textbook develops a general problem-solving strategy that guides you through the steps. To

help you remember the steps of the strategy, they are called *Conceptualize*, *Categorize*, *Analyze*, and *Finalize*.

General Problem-Solving Strategy

Conceptualize

- The first thing to do when approaching a problem is to *think about* and *understand* the situation. Read the problem several times until you are confident you understand what is being asked. Study carefully any diagrams, graphs, tables, or photographs that accompany the problem. Imagine a movie, running in your mind, of what happens in the problem.

- If a diagram is not provided, you should almost always make a quick drawing of the situation. Indicate any known values, perhaps in a table or directly on your sketch.

- Now focus on what algebraic or numerical information is given in the problem. In the problem statement, look for key phrases such as "starts from at rest" ($v_i = 0$), "stops" ($v_f = 0$), or "freely falls" ($a_y = -g = -9.80 \text{ m/s}^2$). Key words can help simplify the problem.

- Next focus on the expected result of solving the problem. Exactly what is the question asking? Will the final result be numerical or algebraic? If it is numerical, what units will it have? If it is algebraic, what symbols will appear in it?

- Incorporate information from your own experiences and common sense. What should a reasonable answer look like? What should its order of magnitude be? You wouldn't expect to calculate the speed of an automobile to be $5 \times 10^6 \text{ m/s}$.

Categorize

- Once you have a really good idea of what the problem is about, you need to *simplify* the problem. Remove the details that are not important to the solution. For example, you can often model a moving object as a particle. Key words should tell you whether you can ignore air resistance or friction between a sliding object and a surface.

- Once the problem is simplified, it is important to *categorize* the problem. How does it fit into a framework of ideas that you construct to understand the world? Is it a simple *plug-in problem*, such that numbers can be simply substituted into a definition? If so, the problem is likely to be finished when this substitution is done.

If not, you face what we can call an *analysis problem*—the situation must be analyzed more deeply to reach a solution.

- If it is an analysis problem, it needs to be categorized further. Have you seen this type of problem before? Does it fall into the growing list of types of problems that you have solved previously? Being able to classify a problem can make it much easier to lay out a plan to solve it. For example, if your simplification shows that the problem can be treated as a particle moving under constant acceleration and you have already solved such a problem, the solution to the new problem follows a similar pattern.

Analyze

- Now, you need to analyze the problem and strive for a mathematical solution. Because you have already categorized the problem, it should not be too difficult to select relevant equations that apply to the type of situation in the problem.

- Use algebra (and calculus, if necessary) to solve symbolically for the unknown variable in terms of what is given. Substitute in the appropriate numbers, calculate the result, and round it to the proper number of significant figures.

Finalize

- This final step is the most important part. Examine your numerical answer. Does it have the correct units? Does it meet your expectations from your conceptualization of the problem? What about the algebraic form of the result—before you substituted numerical values? Does it make sense? Try looking at the variables in it to see whether the answer would change in a physically meaningful way if they were drastically increased or decreased or even became zero. Looking at limiting cases to see whether they yield expected values is a very useful way to make sure that you are obtaining reasonable results.

- Think about how this problem compares with others you have done. How was it similar? In what critical ways did it differ? Why was this problem assigned? You should have learned something by doing it. Can you figure out what? Can you use your solution to expand, strengthen, or otherwise improve your framework of ideas? If it is a new category of problem, be sure you understand it so that you can use it as a model for solving future problems in the same category.

When solving complex problems, you may need to identify a series of sub-problems and apply the problem-solving strategy to each. For very simple problems, you probably don't need this whole strategy. But when you are looking at a problem and you don't know what to do next, remember the steps in the strategy and use them as a guide.

Work on problems in this Study Guide yourself and compare your solutions with ours. Your solution does not have to look just like the one presented here. A problem can sometimes be solved in different ways, starting from different principles. If you wonder about the validity of an alternative approach, ask your instructor.

5. We suggest that you use this Study Guide to review the material covered in the text, and as a guide in preparing for exams. You can use the sections Notes From Selected Chapter Sections, and Equations and Concepts to focus in on any points which require further study. The main purpose of this Study Guide is to improve upon the efficiency and effectiveness of your study hours and your overall understanding of physical concepts. However, it should not be regarded as a substitute for your textbook or for individual study and practice in problem solving.

Chapter 15

Electric Forces and Electric Fields

NOTES FROM SELECTED CHAPTER SECTIONS

15.1 Properties of Electric Charges

Electric charge has the following important properties:

- There are two kinds of charges (positive and negative) in nature. Unlike charges attract one another, and like charges repel one another.

- The force between charges varies as the inverse square of the distance between them.

- Charge is always conserved.

- Charge comes in discrete packets (quantized) that are integral multiples of the electronic charge, $e = 1.6 \times 10^{-19}$ C.

An object becomes charged electrically due to the gain or loss of electrons. A negatively charged object has an excess of electrons (relative to protons), and a positively charged object has a deficiency of electrons.

15.2 Insulators and Conductors

Conductors are materials in which electrons move freely under the influence of an electric field; insulators are materials that do not readily transport charge.

15.3 Coulomb's Law

Experiments show that an electric force between two charges is:

- Inversely proportional to the square of the distance between the two charges and is along the line joining them.

- Proportional to the product of the magnitudes of the charges,

- Attractive if the charges are of opposite sign and repulsive if the charges have the same sign.

The electric interaction between two charges q_1 and q_2 obeys Newton's third law. The two charges experience forces that are equal in magnitude and opposite in direction, regardless of the relative magnitude of the two charges.

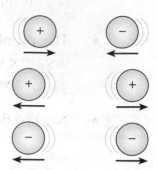

15.4 The Electric Field

An electric field exists at a point if a test charge placed at that point experiences an electric force. The electric field vector $\vec{E}$ at a point in space is placed at that point divided by the magnitude of the test charge q_0. The SI units of the electric field are newtons per coulomb (N/C).

The total electric field at a point, due to a group of charges, equals the vector sum of the individual electric fields at that point due to each of the charges. *This is called the superposition principle.*

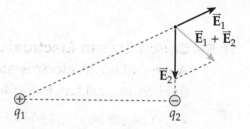

15.5 Electric Field Lines

A convenient aid for visualizing electric field patterns can be obtained by drawing a group of electric field lines, which indicate the direction of the electric field vector at any point. These lines are related to the electric field in any region of space in the following manner:

- The electric field vector $\vec{E}$ is tangent to the electric field line at each point.

- The number of electric field lines per unit area through a surface perpendicular to the lines is proportional to the strength of the electric field in that region. Thus, $\vec{E}$ is large when the field lines are close together and small when they are far apart.

The rules for drawing electric field lines for any charge distribution are as follows:

- The lines must begin on positive charges and terminate on negative charges, or at infinity in the case of an excess of charge.

- The number of lines drawn leaving a positive charge or terminating on a negative charge is proportional to the magnitude of the charge. In the case of an isolated charge, the field lines are evenly distributed around the charge.

- No two field lines can cross each other.

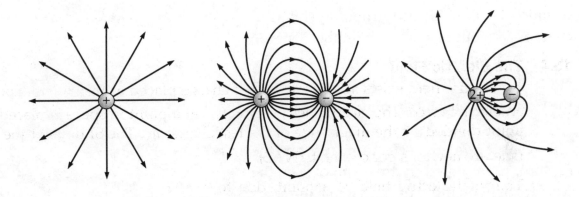

15.6   Conductors in Electrostatic Equilibrium

A conductor in electrostatic equilibrium (no net motion of charge occurs within the conductor) has the following properties:

- The electric field is zero everywhere inside the conductor.

- Any excess charge on an isolated conductor resides entirely on its surface.

- The electric field just outside a charged conductor is perpendicular to the surface of the conductor.

- On an irregularly shaped conductor, the charge per unit area is greatest at locations of sharpest curvature, that is, where the radius of curvature is smallest.

15.9   Electric Flux and Gauss's Law

There is an important general relation between the net electric flux through an imaginary closed surface (often called a gaussian surface) and the charge enclosed by the surface. This is known as Gauss's law and states that the net electric flux through any closed gaussian surface is equal to the net charge inside the surface divided by ϵ_0. Gauss's law is valid for any closed surface; in practical terms, it is useful for calculating the electric field in situations which have a high degree of symmetry in the charge distribution (for example uniformly charged spheres, cylinders, or planes).

EQUATIONS AND CONCEPTS

Coulomb's law gives the magnitude of the electrostatic force between two stationary point charges q_1 and q_2, separated by a distance r. This is also true for spherical charge distributions when r is the distance between their centers. In calculations the approximate value (8.99×10^9 N·m²/C²) of the Coulomb constant, k_e, may be used.

$$F = k_e \frac{|q_1||q_2|}{r^2} \qquad (15.1)$$

$$k_e = 8.987\ 5 \times 10^9 \ \text{N·m}^2/\text{C}^2$$

The **charge on the electron** (or on the proton), represented by the symbol, e, is the fundamental quantity of charge in nature.

$$e = 1.6 \times 10^{-19} \ \text{C}$$

Definition of the electric field, $\vec{E}$, at any point in space is the ratio of electric force per unit charge exerted on a small positive test charge, q_0, placed at the point where the field is to be determined. **The direction of $\vec{E}$ is along the direction of $\vec{F}$.**

$$\vec{E} \equiv \frac{\vec{F}}{q_0} \qquad (15.4)$$

The **electric field of a point charge** has a magnitude given by Eq. 15.6. The direction of the electric field is radially outward from a positive point charge and radially inward toward a negative point charge. *The superposition principle holds when the electric field at a point is due to a number of point charges.*

$$E = k_e \frac{|q|}{r^2} \qquad (15.6)$$

Electric flux through a surface is proportional to the number of electric lines that penetrate the surface. For a plane surface in a uniform field, the flux depends on the angle between the normal to the surface and the direction of the field. Electric flux has SI units of $N \cdot m^2 / C$. In the figure, the electric field is parallel to the normal to the plane surface and therefore, in this case, $\theta = 0$ in Eq.15.9.

$$\Phi_E = EA \cos\theta \qquad (15.9)$$

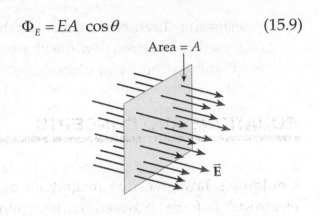

The electric flux through the area, A, is equal to EA

Gauss's law states that the net electric flux, Φ_E, out of a closed surface (gaussian surface) is equal to the net charge enclosed by the surface divided by the constant ϵ_0. *By convention for a closed surface, the flux lines passing into the interior of the volume are negative, and those passing out of the interior of the volume are positive.*

$$EA = \Phi_E = \frac{Q_{inside}}{\epsilon_0} \qquad (15.12)$$

$$Q_{inside} = Q_{positive} - Q_{negative}$$

SUGGESTIONS, SKILLS, AND STRATEGIES

ELECTRIC FORCES AND FIELDS

1. **Units:** When performing calculations using Coulomb's law and the Coulomb constant $k_e = 8.99 \times 10^9 \ N \cdot m^2 / C^2$, charge must be in coulombs and distances in meters. If they are given in other units, you must convert them to SI units.

2. **Applying Coulomb's law to point charges:** Remember to use the magnitude of the charges in Equation 15.1. The direction of the force is found by noting that the forces are repulsive between like charges and attractive between unlike charges. Use the superposition principle properly when dealing with a collection of interacting point charges. When several charges are present, the resultant force on one of them is found by first calculating (magnitude and direction) the force that each of the other charges exerts on it. Then, determine the vector sum of these forces. The magnitude of the force that any charged particle exerts on another is given by Coulomb's law.

3. **Calculating the electric field due to point charges:** The direction of the electric field is the direction of electric force on a positive "test" charge if placed at the point in question. The superposition principle can also be applied to electric fields, which are also vector quantities. To find the total electric field at a given point, first calculate the electric field at the point due to each individual charge. The resultant field at the point is the vector sum of the fields due to the individual charges.

4. **Electric Flux and Gauss's Law:** When using Gauss's law to calculate the electric field in the vicinity of a charge distribution, you must anticipate the expected symmetry of the electric field by considering the shape of the charge distribution. Choose a gaussian surface which has a shape matching the symmetry of the field (spherical surface for a point charge, cylindrical surface for a line of charge, etc.). The gaussian surface must be closed and include the point at which the field is to be calculated.

    An important factor in your choice of a shape should be the ease with which you can calculate the flux through the closed surface. This will be the case when the surface can be divided into regions so that over a particular region either (1) the flux is zero, or (2) the electric field is constant. Consider, for example, a line of charge along the axis of a closed cylindrical surface.

REVIEW CHECKLIST

▷ Use Coulomb's law to determine the net electrostatic force (magnitude and direction) on a point electric charge due to a known distribution of a finite number of point charges.

▷ Calculate the electric field $\bar{E}$ (magnitude and direction) at a specified location in the vicinity of a group of point charges.

▷ Describe the pattern of electric field lines associated with charge distributions such as (i) single point charges, (ii) pairs of charges with like or unlike sign and equal or unequal magnitudes, (iii) charged metallic sphere, etc.

▷ State and justify the conditions for the charge distribution on a conductor in electrostatic equilibrium.

▷ Gauss's law offers a useful technique for calculating the electric field due to symmetric charge distributions (point, sphere, line, cylinder, plane). The electric flux through a closed surface (gaussian surface) is equal to the net charge enclosed by the surface divided by ϵ_0. Remember that lines passing into the surface are negative, and lines passing out of the surface are positive. Be able to identify gaussian surfaces that would be useful in determining the electric field due to each of the symmetric charge distributions mentioned above.

SOLUTIONS TO SELECTED END-OF-CHAPTER PROBLEMS

5.   The nucleus of ^{8}Be, which consists of 4 protons and 4 neutrons, is very unstable and spontaneously breaks into two alpha particles (helium nuclei, each consisting of 2 protons and 2 neutrons). (a) What is the force between the two alpha particles when they are 5.00×10^{-15} m apart, and (b) what will be the magnitude of the acceleration of the alpha particles due to this force? Note that the mass of an alpha particle is 4.002 6 u.

Solution

(a)   Treat the two alpha particles, resulting from the beryllium decay, as point charges with

$$q_1 = q_2 = +2e = 2\left(1.60 \times 10^{-19} \text{ C}\right) = 3.20 \times 10^{-19} \text{ C}$$

When separated by a distance $r = 5.00 \times 10^{-15}$ m, these positive charges repel each other with a force whose magnitude is given by Coulomb's law:

$$F_e = \frac{k_e q_1 q_2}{r^2} = \left(8.99 \times 10^9 \text{ N} \cdot \text{m}^2/\text{C}^2\right)\frac{\left(3.20 \times 10^{-19} \text{ C}\right)^2}{\left(5.00 \times 10^{-15} \text{ m}\right)^2} = 36.8 \text{ N} \qquad \Diamond$$

(b)   The mass of an alpha particle is $\qquad\qquad m = 4.002\,6$ u

where the unified mass unit is given by $\qquad\qquad 1\text{u} = 1.66 \times 10^{-27}$ kg

Newton's second law then gives the magnitude of the acceleration of each alpha particle as

$$a = \frac{F_e}{m} = \frac{36.8 \text{ N}}{4.002\,6\left(1.66 \times 10^{-27} \text{ kg}\right)} = 5.54 \times 10^{27} \text{ m/s}^2 \qquad \Diamond$$

9. Two identical conducting spheres are placed with their centers 0.30 m apart. One is given a charge of 12×10^{-9} C, the other a charge of -18×10^{-9} C. (a) Find the electrostatic force exerted on one sphere by the other. (b) The spheres are connected by a conducting wire. Find the electrostatic force between the two after equilibrium is reached.

Solution

(a) Assuming the charge on each sphere is uniformly distributed, the spheres behave as if all the charge on them was concentrated as point charges at their centers. Coulomb's law then gives the magnitude of the force one sphere exerts on the other as

$$F_e = k_e \frac{q_1 |q_2|}{r^2} = \left(8.99 \times 10^9\right) \frac{\left(12 \times 10^{-9} \text{ C}\right)\left(18 \times 10^{-9} \text{ C}\right)}{\left(0.30 \text{ m}\right)^2} = 2.2 \times 10^{-5} \text{ N} \qquad \Diamond$$

(b) When a conducting wire connects the two spheres, charge will flow from one sphere to the other until equilibrium is established. Since the two spheres are identical, the total excess charge on the two spheres will be divided equally between the two spheres when equilibrium occurs. Therefore, each sphere will then have a net charge of

$$q_f = \frac{q_1 + q_2}{2} = \frac{12 \times 10^{-9} \text{ C} - 18 \times 10^{-9} \text{ C}}{2} = -3.0 \times 10^{-9} \text{ C}$$

After equilibrium is reached, the two spheres will repel each other with forces of magnitude

$$F'_e = k_e \frac{q_f q_f}{r^2} = \left(8.99 \times 10^9\right) \frac{\left(-3.0 \times 10^{-9} \text{ C}\right)\left(-3.0 \times 10^{-9} \text{ C}\right)}{\left(0.30 \text{ m}\right)^2} = 9.0 \times 10^{-7} \text{ N} \qquad \Diamond$$

15.   Two small metallic spheres, each of mass 0.20 g, are suspended as pendulums by light strings from a common point as shown in Figure P15.15. The spheres are given the same electric charge, and it is found that they come to equilibrium when each string is at an angle of 5.0° with the vertical. If each string is 30.0 cm long, what is the magnitude of the charge on each sphere?

Solution

Consider the sketch at the right showing the two spheres and the forces acting on one. When equilibrium is reached, we have

$$\Sigma F_y = 0 \;\Rightarrow\; T\cos\theta = mg \quad \text{or} \quad T = \frac{mg}{\cos\theta}$$

and

$$\Sigma F_x = 0 \;\Rightarrow\; F_e = T\sin\theta = \left(\frac{mg}{\cos\theta}\right)\sin\theta$$

or $F_e = mg\tan\theta$

But, from Coulomb's law, we know that

$$F_e = k_e\frac{q_1 q_2}{r^2} = k_e\frac{q^2}{r^2} \quad \text{where } r = 2(L\sin\theta)$$

Thus, $k_e\dfrac{q^2}{r^2} = mg\tan\theta$ or $q = \sqrt{\dfrac{mgr^2\tan\theta}{k_e}} = \sqrt{\dfrac{4mgL^2\sin^2\theta\tan\theta}{k_e}}$

If $m = 0.20\times10^{-3}$ kg, $L = 0.300$ m, and $\theta = 5.0°$, the magnitude of the charge on each sphere is

$$q = \sqrt{\frac{4\left(0.20\times10^{-3}\ \text{kg}\right)\left(9.8\ \text{m/s}^2\right)\left(0.300\ \text{m}\right)^2\sin^2\left(5.0°\right)\tan\left(5.0°\right)}{8.99\times10^9\ \text{N}\cdot\text{m}^2/\text{C}^2}}$$

or $q = 7.2\times10^{-9}$ C $= 7.2$ nC   ◊

19. An airplane is flying through a thundercloud at a height of 2 000 m. (This is a very dangerous thing to do because of updrafts, turbulence, and the possibility of electric discharge.) If there are charge concentrations of +40.0 C at a height of 3 000 m within the cloud and –40.0 C at a height of 1 000 m, what is the electric field $\vec{E}$ at the aircraft?

Solution

The resultant electric field at the plane's location consists of two contributions, one due to each charge concentration. If we treat the concentrations as point charges, each makes a contribution of magnitude $E = k_e |q|/r^2$ to the total electric field. Here, q is the total charge of the concentration, and r is the distance from the concentration to the plane.

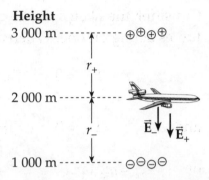

At the 2 000-m height, the contribution by the charge located at 3 000 m is directed **downward**, <u>away from</u> that concentration of positive charge. Its magnitude is

$$E_+ = k_e \frac{|q_+|}{r_+^2} = (8.99 \times 10^9 \ \text{N} \cdot \text{m}^2/\text{C}^2) \frac{(40.0 \ \text{C})}{(1\,000 \ \text{m})^2} = 3.60 \times 10^5 \ \text{N/C}$$

At the plane's location, the contribution by the charge at 1 000 m altitude is directed **downward** (<u>toward</u> that concentration of negative charge) and has magnitude

$$E_- = k_e \frac{|q_-|}{r_-^2} = (8.99 \times 10^9 \ \text{N} \cdot \text{m}^2/\text{C}^2) \frac{(40.0 \ \text{C})}{(1\,000 \ \text{m})^2} = 3.60 \times 10^5 \ \text{N/C}$$

The resultant field is then

$$\vec{E} = \vec{E}_+ + \vec{E}_- = 7.20 \times 10^5 \ \text{N/C} \quad \text{(downward)}$$

◊

23.   A proton accelerates from rest in a uniform electric field of 640 N/C. At some later time, its speed is 1.20×10^6 m/s. (a) Find the magnitude of the acceleration of the proton. (b) How long does it take the proton to reach this speed? (c) How far has it moved in that interval? (d) What is its kinetic energy at the later time?

Solution

(a)   The force exerted on the proton by the electric field has magnitude $F_e = qE$. The charge and mass of the proton are:

$$q = +e = 1.60 \times 10^{-19} \text{ C}$$

and

$$m_p = 1.673 \times 10^{-27} \text{ kg}$$

Newton's second law then gives the acceleration as

$$a = \frac{F_e}{m_p} = \frac{eE}{m_p} = \frac{\left(1.60 \times 10^{-19} \text{ C}\right)\left(640 \text{ N/C}\right)}{1.673 \times 10^{-27} \text{ kg}} = 6.12 \times 10^{10} \text{ m/s}^2 \qquad \Diamond$$

(b)   The force, and hence the acceleration, is constant in a uniform electric field. Therefore, the time required to reach the final speed is

$$t = \frac{v_f - v_i}{a} = \frac{1.20 \times 10^6 \text{ m/s} - 0}{6.12 \times 10^{10} \text{ m/s}^2} = 1.96 \times 10^{-5} \text{ s} = 19.6 \text{ } \mu s \qquad \Diamond$$

(c)   The uniformly accelerated motion equation $v_f^2 = v_i^2 + 2a(\Delta x)$ gives the displacement of the proton during this time as

$$\Delta x = \frac{v_f^2 - v_i^2}{2a} = \frac{\left(1.20 \times 10^6 \text{ m/s}\right)^2 - 0}{2\left(6.12 \times 10^{10} \text{ m/s}^2\right)} = 11.8 \text{ m} \qquad \Diamond$$

(d)   The kinetic energy of the proton at the end of this time interval is

$$KE_f = \frac{1}{2} m_p v_f^2 = \frac{1}{2}\left(1.673 \times 10^{-27} \text{ kg}\right)\left(1.20 \times 10^6 \text{ m/s}\right)^2 = 1.20 \times 10^{-15} \text{ J} \qquad \Diamond$$

This result can also be obtained from the work-energy theorem $W_{net} = KE_f - KE_i$, where $W_{net} = F_e(\Delta x) = eE(\Delta x)$, as follows:

$$KE_f = W_{net} + KE_i = eE(\Delta x) + 0$$

$$= \left(1.60 \times 10^{-19} \text{ C}\right)\left(640 \text{ N/C}\right)\left(11.8 \text{ m}\right) = 1.20 \times 10^{-15} \text{ J} \qquad \Diamond$$

31. Two point charges are a small distance apart. (a) Sketch the electric field lines for the two if one has a charge four times that of the other and both charges are positive. (b) Repeat for the case in which both charges are negative.

Solution

In sketching the electric field patterns for given charge distributions, there are several points one should keep in mind.

Some of these are:

(1) When point charges are isolated in space, the field lines exhibit radial symmetry. That is, they are uniformly spaced around the charge and either radially outward away from the charge or radially inward toward the charge. When in the presence of other charges, the field lines will still be nearly radial at points very close to the charge but will become distorted farther out.

(2) Field lines originate on positive charges and terminate on negative charges.

(3) The number of field lines one should draw leaving a positive charge or approaching a negative charge is proportional to the charge. Therefore, the number of lines leaving (or approaching) a charge of magnitude $4q$ should be four times the number leaving (or approaching) a charge of magnitude q.

(4) Field lines never cross each other.

(a) In the drawing below for two point charges of magnitudes q and $4q$, located near each other, not all of the lines leaving the charge $4q$ are shown for the sake of clarity in the drawing. ◊

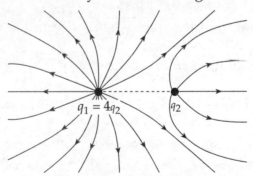

(b) If the charges q_1 and q_2 were negative instead of positive, the pattern of field lines would be as shown above with the exception that the arrows would point in the opposite direction along the lines (that is, in toward the negative charges). ◊

35.   If the electric field strength in air exceeds 3.0×10^6 N/C, the air becomes a conductor. Using this fact, determine the maximum amount of charge that can be carried by a metal sphere 2.0 m in radius. (See the hint in Problem 34.)

Solution

In electrostatic equilibrium, the excess charge on a conducting sphere is uniformly distributed over the outer surface of that sphere. Also, at all points **outside** a spherically symmetric charge distribution, the electric field is identical to that which would exist if the total charge was concentrated as a point charge located at the center of the sphere.

Thus, the electric field strength at any point outside the charged metal sphere is given by

$$E = k_e Q/r^2$$

where Q is the net charge on the sphere, and r is the distance to this point from the center of the sphere.

Notice that the field has the greatest strength where r has its minimum allowed value. Since the observation point must be **outside** the spherical charge distribution, the minimum allowed value is $r = R$ where R is the radius of the sphere.

Therefore, the field is strongest just outside the surface of the sphere, and as the charge on the sphere is increased, electrical breakdown will begin to occur (that is, the air will become conducting) when the electric field at the surface of the sphere reaches 3.0×10^6 N/m.

The maximum charge that the sphere may have is found from

$$E\big|_{r=R} = \frac{k_e Q_{max}}{R^2} = 3.0 \times 10^6 \text{ N/m} \text{, or } Q_{max} = \frac{\left(3.0 \times 10^6 \text{ N/m}\right) R^2}{k_e}$$

If $R = 2.0$ m, $Q_{max} = \dfrac{\left(3.0 \times 10^6 \text{ N/m}\right)\left(2.0 \text{ m}\right)^2}{8.99 \times 10^9 \text{ N} \cdot \text{m}^2/\text{C}^2} = 1.3 \times 10^{-3}$ C   ◊

39.   An electric field of intensity 3.50 kN/C is applied along the *x*-axis. Calculate the electric flux through a rectangular plane 0.350 m wide and 0.700 m long if (a) the plane is parallel to the *yz*-plane; (b) the plane is parallel to the *xy*-plane; and (c) the plane contains the *y*-axis, and its normal makes an angle of 40.0° with the *x*-axis.

Solution

The planar section of interest has area

$$A = length \times width = (0.700 \text{ m}) \times (0.350 \text{ m}) = 0.245 \text{ m}^2$$

When this area is in a uniform electric field of magnitude *E*, the flux through it will be

$$\Phi_E = EA\cos\theta$$

where θ is the angle that the perpendicular or "normal" to the area makes with the direction of the electric field (the positive *x*-axis in this case).

(a)   If the area is parallel to the *yz*-plane, then its perpendicular is parallel to the *x*-axis, and hence, to the field. In this case, $\theta = 0°$, and the flux is

$$\Phi_E = (3.50 \times 10^3 \text{ N/C})(0.245 \text{ m}^2)\cos 0° = 858 \text{ N} \cdot \text{m}^2/\text{C} \qquad \Diamond$$

(b)   When the area is parallel to the *xy*-plane, its normal is parallel to the *z*-axis, and hence, makes an angle of 90.0° with the direction of the field. The flux is then

$$\Phi_E = (3.50 \times 10^3 \text{ N/C})(0.245 \text{ m}^2)\cos 90.0° = 0 \qquad \Diamond$$

If the plane of the area contains the *y*-axis and its normal makes an angle of 40.0° with the *x*-axis, then $\theta = 40.0°$ and the flux is

$$\Phi_E = (3.50 \times 10^3 \text{ N/C})(0.245 \text{ m}^2)\cos 40.0° = 657 \text{ N} \cdot \text{m}^2/\text{C} \qquad \Diamond$$

43. A point charge q is located at the center of a spherical shell of radius a that has a charge $-q$ uniformly distributed on its surface. Find the electric field (a) for all points outside the spherical shell and (b) for a point inside the shell a distance r from the center.

Solution

With the positive point change at the center of the spherical shell, and the negative charge uniformly distributed on the surface of the shell, the charge distribution has complete spherical symmetry. Thus, anywhere that an electric field due to this distribution exists, that field will be in the radial direction, and its magnitude will depend only on the distance r from the center of the spherical shell.

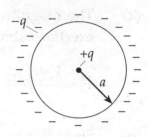

To take advantage of the symmetry known to exist in the field, we shall employ a spherical gaussian surface that is concentric with the spherical shell. The flux through this surface is then

$$\Phi_E = EA\cos\theta = E\left(4\pi r^2\right)\cos 0° = 4\pi r^2 E$$

where r is the radius of the gaussian surface, and E is the magnitude of the electric field at all points on this surface.

Gauss's law also tells us that the flux through our gaussian surface is equal to the net charge enclosed by the surface divided by ϵ_0. Thus, we have

$$\Phi_E = 4\pi r^2 E = Q_{net}/\epsilon_0 \qquad \text{or} \qquad E = Q_{net}/4\pi \epsilon_0 r^2$$

(a) If our gaussian surface is outside the spherical shell, $r > a$, then $Q_{net} = +q - q = 0$. Thus, the electric field at all points outside the spherical shell has zero magnitude. ◊

(b) If the radius of our gaussian surface is less than that of the shell, $0 < r < a$, then $Q_{net} = +q$, and the electric field at points inside the shell is

$$\vec{E} = q/4\pi \epsilon_0 r^2 \quad \text{radial outward (away from the positive charge)} \qquad ◊$$

48. In the Bohr theory of the hydrogen atom, an electron moves in a circular orbit about a proton. The radius of the orbit is 0.53×10^{-10} m. (a) Find the electrostatic force acting on each particle. (b) If this force causes the centripetal acceleration of the electron, what is the speed of the electron?

Solution

(a) The magnitude of the electrostatic force of attraction the electron and proton exert on each other is given by Coulomb's law as

$$F_e = \frac{k_e |q_1||q_2|}{r^2} = \frac{k_e e^2}{r^2} = \frac{\left(8.99 \times 10^9 \ \text{N} \cdot \text{m}^2/\text{C}^2\right)\left(1.60 \times 10^{-19} \ \text{C}\right)^2}{\left(0.53 \times 10^{-10} \ \text{m}\right)^2} = 8.2 \times 10^{-8} \ \text{N} \qquad \Diamond$$

(b) Since the proton is much more massive than an electron, the proton may be considered to remain at rest as the electron moves in an orbit around it. A particle moving at speed v in a circular path of radius r has an acceleration of magnitude $a_c = v^2/r$ directed toward the center of the path. If the particle has mass m_e, the force required to produce this centripetal acceleration is

$$F = m_e a_c = \frac{m_e v^2}{r}$$

In the case of the hydrogen atom, the electrostatic force which the proton exerts on the electron supplies this required force. Thus, the speed of the electron when it moves in an orbit of radius $r = 0.53 \times 10^{-10}$ m must be

$$v = \sqrt{\frac{F_e r}{m_e}} = \sqrt{\frac{\left(8.2 \times 10^{-8} \ \text{N}\right)\left(0.53 \times 10^{-10} \ \text{m}\right)}{9.11 \times 10^{-31} \ \text{kg}}} = 2.2 \times 10^6 \ \text{m/s} \qquad \Diamond$$

51.   (a) Two identical point charges $+q$ are located on the y-axis at $y = +a$ and $y = -a$. What is the electric field along the x-axis at $x = b$? (b) A circular ring of charge of radius a has a total positive charge Q distributed uniformly around it. The ring is in the $x = 0$ plane with its center at the origin. What is the electric field along the x-axis at $x = b$ due to the ring of charge? [Hint: Consider the charge Q to consist of many pairs of identical point charges positioned at the ends of diameters of the ring.]

Solution

(a)   Any point on the x-axis is equidistant from the two identical point charges. Thus, the contributions by the two charges to the resultant field at this point have equal magnitudes given by

$$E_1 = E_2 = \frac{k_e q}{r^2}$$

The components of the resultant field are

$$E_y = E_{1y} - E_{2y}: \qquad E_y = \left(\frac{k_e q}{r^2}\right)\sin\theta - \left(\frac{k_e q}{r^2}\right)\sin\theta = 0$$

and $E_x = E_{1x} + E_{2x}:$

$$E_x = \left(\frac{k_e q}{r^2}\right)\cos\theta + \left(\frac{k_e q}{r^2}\right)\cos\theta = \left[\frac{k_e(2q)}{r^2}\right]\cos\theta$$

Since $r = \sqrt{a^2 + b^2}$ and $\cos\theta = b/r$

the net x-component becomes $E_x = \dfrac{k_e(2q)b}{r^3} = \dfrac{k_e b(2q)}{\left(a^2 + b^2\right)^{3/2}}$

The resultant electric field on the axis at $x = b$ is:

$$\vec{E}_R = \frac{k_e b(2q)}{\left(a^2 + b^2\right)^{3/2}} \quad \text{in the +x direction} \qquad \diamond$$

(b)   Note that the result of part (a) may be written as $\vec{E}_R = \dfrac{k_e b(Q)}{\left(a^2 + b^2\right)^{3/2}}$

where $Q = 2q$ is the total charge in the charge distribution generating the field.

In the case of a uniformly charged circular ring, consider the ring to consist of many pairs of identical charges uniformly spaced around the ring. Each pair of charges has a total charge Q_i. At a point on the axis of the ring, this pair of charges generates an electric field contribution that is parallel to the axis and has magnitude

$$E_i = \frac{k_e \, b Q_i}{\left(a^2 + b^2\right)^{3/2}}$$

The resultant electric field of the ring is the summation of the contributions by all such pairs of charges, or

$$E_R = \Sigma E_i = \left[\frac{k_e b}{\left(a^2 + b^2\right)^{3/2}}\right]\Sigma Q_i = \frac{k_e b Q}{\left(a^2 + b^2\right)^{3/2}}$$

where $Q = \Sigma Q_i$ is the total charge on the ring.

Thus, the electric field at a point on the axis of the ring (radius a), and distance b from the plane of the ring, is directed parallel to the axis and away from the ring. Its magnitude is

$$\left|\vec{\mathbf{E}}_R\right| = \frac{k_e \, b Q}{\left(a^2 + b^2\right)^{3/2}}$$

◊

57. Two 2.0-g spheres are suspended by 10.0-cm-long light strings (Fig. P15.57). A uniform electric field is applied in the x-direction. If the spheres have charges of -5.0×10^{-8} C and $+5.0 \times 10^{-8}$ C, determine the electric field intensity that enables the spheres to be in equilibrium at $\theta = 10°$.

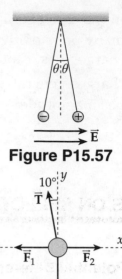

Figure P15.57

Solution

The sketch at the right gives a free-body diagram of the positively charged sphere. Here, $F_1 = k_e |q|^2 / r^2$ is the attractive force exerted by the negatively charged sphere and $F_2 = qE$ is exerted by the electric field.

Since the sphere is in equilibrium, we know that

$$\Sigma F_y = T\cos 10° - mg = 0 \quad \text{or} \quad T = \frac{mg}{\cos 10°} \qquad \text{[1]}$$

Also, $\Sigma F_x = F_2 - F_1 - T\sin 10° = 0 \quad \text{or} \quad F_2 = F_1 + T\sin 10°$

Substituting for F_1 and F_2, and using Equation 1 to eliminate the tension T gives

$$qE = \frac{k_e |q|^2}{r^2} + mg\tan 10° \qquad \text{[2]}$$

At equilibrium, the distance between the two spheres is $r = 2(L\sin 10°)$, where L is the length of the string holding each sphere. Thus, Equation 2 becomes

$$E = \frac{k_e |q|}{4(L\sin 10°)^2} + \frac{mg\tan 10°}{q}$$

$$= \frac{(8.99 \times 10^9 \ \text{N} \cdot \text{m}^2/\text{C}^2)(5.0 \times 10^{-8} \ \text{C})}{4[(0.100 \ \text{m})\sin 10°]^2} + \frac{(2.0 \times 10^{-3} \ \text{kg})(9.8 \ \text{m/s}^2)\tan 10°}{(5.0 \times 10^{-8} \ \text{C})}$$

or the required electric field strength is $\quad E = 4.4 \times 10^5 \ \text{N/C}$ ◊

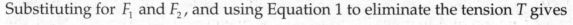

Chapter 16

Electrical Energy and Capacitance

NOTES ON SELECTED CHAPTER SECTIONS

16.1 Potential Difference and Electric Potential

The electrostatic force is conservative; therefore, it is possible to define an electric potential energy function associated with this force. The potential difference between two points is proportional to the change in potential energy of a charge as it moves between the two points.

Electrical potential difference (a scalar quantity) is the work done to move a charge from point A to point B divided by the magnitude of the charge. Thus, the SI units of potential are joules per coulomb and are called volts (V).

A positive charge gains electrical potential energy when it is moved opposite to the direction of the electric field. When a positive charge is placed in an electric field, it moves in the direction of the field, from a point of high potential to a point of lower potential.

16. 2 Electric Potential and Potential Energy Due to Point Charges

In electric circuits, a point of zero potential is often defined by grounding (connecting to Earth) some point in the circuit. In the case of a point charge, the point of zero potential is taken to be at an infinite distance from the charge. The electric potential at a given point in space due to a point charge q depends only on the value of the charge and the distance, r, from the charge to the specified point in space. An electric potential can exist at a point in space whether or not a test charge exists at that point.

Electric potential is a scalar quantity. When using the superposition principle to determine the value of the electric potential at a point due to several point charges, the algebraic sum of the individual electric potentials must be used. *Be careful not to confuse electric potential difference with electric potential energy.*

Electric potential is a scalar property of the region (electric field) surrounding an electric charge. It does not depend on the presence of a test charge in the field.

Electric potential energy is a characteristic of a charge-field system. It is due to the interaction of the field and a charge located within the field.

In the figure to the right an electric field (due to the presence of $+Q$) is present in the region surrounding the charge. This field is due to a point charge and has the properties described in Chapter 15.

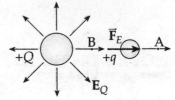

Refer to the figure as you consider the following statements about electric potential difference and electric potential energy.

- Electric field lines are directed along outward radials from the charge Q. Points "A" and "B" are shown along one of the field lines.

- If a positive test charge q is placed at some point between A and B, the electric field will exert a force $\vec{F}_E = q\vec{E}$ on q directed along the field line and away from Q.

- If the test charge q is moved from point A to point B (where the potential is greater), the electric field will do work, $W = -q\Delta V = -q(V_B - V_A)$, and the electric potential energy of the two-charge system (Q and q) increases by an amount equal to $q\Delta V$. *The electric potential energy of a positive charge increases when the charge moves to a point closer to another isolated positive point charge (or along a direction opposite the direction of the electric field).*

- *Since the electrostatic force is a conservative force, the work done by the electric field as the test charge q moves from point A to point B will be the same regardless of the actual path taken.*

- You should consider the consequences of changing either or both Q and q from a positive to a negative charge.

16.3   Potentials and Charged Conductors

16.4   Equipotential Surfaces

No work is required to move a charge between two points that are at the same potential. That is, $W = 0$ when $V_B = V_A$.

The electric potential is a constant everywhere on the surface of a charged conductor in electrostatic equilibrium.

The electric potential is constant everywhere inside a conductor, and equal to its value at the surface. This is true because the electric field is zero inside a conductor and no work is required to move a charge between two points inside the conductor.

The electron volt (eV) is defined as the energy that an electron (or proton) gains (or loses) when accelerated through a potential difference of 1 V.

A surface on which every point has the same value of electric potential is an equipotential surface. The electric field at every point on an equipotential surface is perpendicular to the surface; because a parallel component of $\vec{E}$ would imply work would be required to move a charge between two points on the surface.

16.6   Capacitance

A capacitor is a device consisting of a pair of conductors (plates) separated by insulating material. A charged capacitor acts as a storehouse of charge and energy that can be reclaimed when needed for a specific application. The capacitance of a capacitor depends on the physical characteristics of the device (size, shape, plate separation, and the nature of the dielectric medium filling the region between the plates).

The capacitance, C, of a capacitor is defined as the ratio of the magnitude of the charge on either conductor to the magnitude of the potential difference between the conductors. Capacitance has SI units of coulombs per volt, called **farads** (F). The farad is a very large unit of capacitance. In practice, most capacitors have capacitances ranging from picofarads to microfarads.

16.8   Combinations of Capacitors

Two or more capacitors can be connected in a circuit in several possible combinations. For example, three capacitors (each assumed to have the same value of capacitance, C) can be combined as illustrated in the figure at right to achieve four different values of equivalent capacitance.

(a)   all in parallel

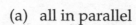 $C_{eq} = 3C$

(b)   all in series

$C_{eq} = \frac{1}{3}C$

(c)   2 in series, in parallel with a third

$C_{eq} = \frac{3}{2}C$

(d)   2 in parallel, in series with a third

$C_{eq} = \frac{2}{3}C$

Can you determine how many different values of equivalent capacitance could be achieved by combining the three capacitors if they each had a different capacitance, C_1, C_2, and C_3?

Note in the figure above that when a group of capacitors is connected in series, they are arranged "end-to-end," and only adjacent capacitors have a circuit point in common. When connected in parallel, each capacitor in the group has two circuit points in common with each of the others.

For capacitors connected in series:

- The reciprocal of the equivalent capacitance equals the sum of the reciprocals of the individual capacitances. See Equation 16.15.

- The potential difference across the series group equals the sum of the potential differences across the individual capacitors.

- Each capacitor of a group in series has the same charge (even if their capacitances are not the same), and this is also the value of the total charge on the series group.

For capacitors connected in parallel:

- The equivalent capacitance equals the sum of the individual capacitances. See Equation 16.12.

- The potential difference across each capacitor has the same value.

- The total charge on the parallel group equals the sum of the charges on the individual capacitors.

16.9 Energy Stored in a Charged Capacitor

The energy stored in a given charged capacitor is proportional to the square of the potential difference between the plates separated by an insulator called a dielectric.

Each capacitor has a limiting voltage that depends on the capacitor's physical characteristics. When the potential difference between the plates of a capacitor exceeds that limiting voltage, a discharge will occur through the insulating material between the two plates. This is known as electrical breakdown. Therefore, the maximum energy which can be stored in a capacitor is limited by the breakdown voltage.

16.10 Capacitors With Dielectrics

A dielectric is an insulating material, such as rubber, glass, waxed paper, or air. When a dielectric is inserted between the plates of a capacitor, the capacitance increases. If the dielectric completely fills the space between the plates, the capacitance is multiplied by a dimensionless factor κ, called the dielectric constant. The dielectric constant is a property of the dielectric material and has a value of 1.000 for a vacuum.

The smallest plate separation for a capacitor is limited by the electric discharge that can occur through the dielectric material separating the plates. For any given plate separation, there is a maximum electric field that can be produced in the dielectric before it breaks down and begins to conduct. This maximum electric field is called the dielectric strength and has SI units of V/m.

EQUATIONS AND CONCEPTS

The **change in electric potential energy** of an electric charge in moving between two points in an electric field is equal to the negative of the work done by the electric force.

$$\Delta PE = -W_{AB} = -(qE_x)\Delta x \qquad (16.1)$$

When the **electric field is uniform**, the change in electric potential energy can be expressed in terms of the charge, the magnitude of the field, and the distance moved parallel to the direction of the field.

$$\Delta PE = -qEd$$
(in a uniform electric field)

The **electric potential difference** between points A and B is defined as the change in electric potential energy per unit charge as a charge is moved from point A to point B. The SI unit of electric potential is the volt, V.

$$\Delta V \equiv V_B - V_A = \frac{\Delta PE}{q} \qquad (16.2)$$

$$1\,V \equiv 1\,J/C$$

The **potential difference in a uniform field** (constant in magnitude and direction depends only on the displacement Δx in a direction parallel to the field. *Electric field lines always point along the direction of decreasing potential.*

$$\Delta V = -\bar{E}_x \Delta x \qquad (16.3)$$

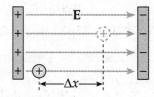

The **electric potential due to a point charge** is inversely proportional to the distance from the charge. *This equation assumes that the potential at infinity is zero. Remember that electric potential is a scalar and the correct algebraic sign for the charge, q, must be used in Equation 16.4.*

$$V = k_e \frac{q}{r} \qquad (16.4)$$

The **total electric potential due to several point charges** is the algebraic sum of the potentials due to the individual charges.

$$V = k_e \sum_i \frac{q_i}{r_i}$$

The potential energy of a pair of charges separated by a distance, *r*, represents the minimum work required to assemble the charges from an infinite separation. *The potential energy of the two charges is positive if the two charges have the same sign; and it is negative if the two charges are of opposite sign.*

$$PE = k_e \frac{q_1 q_2}{r} \tag{16.5}$$

The work required to move a charge *q* from point A to point B in an electric field depends on the magnitude and sign of the charge. *Zero work is required to move a charge between two points that are at the same potential.*

$$W = -q(V_B - V_A) \tag{16.6}$$

Recall that for a conductor in **electrostatic equilibrium**:

Charged conductors in electrostatic equilibrium

- All excess charge resides on the surface.

- E = 0 at all interior points.

- E is perpendicular to the surface on the outside.

- Electric potential is constant in the interior and equal to its value on the surface.

The capacitance, *C*, of a capacitor is defined as the ratio of the charge on either plate (conductor) to the potential difference between the plates (*C* is always positive). The SI unit of capacitance is the farad (F).

$$C \equiv \frac{Q}{\Delta V} \tag{16.8}$$

$1\ \text{F} \equiv 1\ \text{C/V}$

$1\ \text{microfarad} = 10\text{–}6\ \text{F}$

$1\ \text{picofarad} = 10^{-12}\ \text{F}$

The capacitance of an air-filled parallel plate capacitor is proportional to the area of the plates and inversely proportional to the separation of the plates.

$$C = \epsilon_0 \frac{A}{d} \tag{16.9}$$

The value for the **permittivity constant** for free space expressed in SI units.

$$\epsilon_0 = \frac{1}{4\pi k_e} = 8.85 \times 10^{-12}\ \text{C}^2/\text{N} \cdot \text{m}^2$$

For a **parallel combination of capacitors**:

- equivalent capacitance equals the sum of the individual capacitances

$$C_{eq} = C_1 + C_2 + C_3 + \ldots \qquad (16.12)$$

- the total charge is the sum of the charges on the individual capacitors

$$Q_{total} = Q_1 + Q_2 + Q_3 + \ldots$$

- the potential difference is the same for each capacitor

$$\Delta V_{total} = \Delta V_1 = \Delta V_2 = \Delta V_3 = \ldots$$

For **series combination of capacitors**:

- the reciprocal of the equivalent capacitance is equal to the sum of the reciprocals of the individual capacitances

$$\frac{1}{C_{eq}} = \frac{1}{C_1} + \frac{1}{C_2} + \frac{1}{C_3} + \ldots \qquad (16.15)$$

- the potential difference across the group equals the sum of the potential differences of each capacitor

$$\Delta V = \Delta V_1 + \Delta V_2 + \Delta V_3 + \ldots$$

- each capacitor has the same charge

$$Q_{total} = Q_1 = Q_2 = Q_3 = \ldots$$

The **electrostatic potential energy** stored in the electric field of a charged capacitor is equal to the work done by a battery (or other source of emf) in charging the capacitor from $q = 0$ to $q = Q$.

$$W = \tfrac{1}{2} Q \Delta V \qquad (16.16)$$

$$\text{Energy stored} = \tfrac{1}{2} C (\Delta V)^2 = \frac{Q^2}{2C} \qquad (16.17)$$

When a **dielectric** (insulating material) is inserted between the plates of a capacitor, the capacitance of the device increases by a factor κ, the dielectric constant. *Eq.16.19 applies in the case of a parallel- plate capacitor.*

$$C = \kappa \epsilon_0 \left(\frac{A}{d} \right) \qquad (16.19)$$

SUGGESTIONS, SKILLS, AND STRATEGIES

A STRATEGY FOR PROBLEMS INVOLVING ELECTRIC POTENTIAL

1. When working problems involving electric potential, remember that electric potential is a scalar quantity (rather than a vector quantity like the electric field), so there are no components to worry about. Therefore, when using the superposition principle to evaluate the electric potential at a point due to a system of point

charges, you simply take the algebraic sum of the electric potentials due to each charge. However, you must keep track of signs. The electric potential due to each positive charge is positive; the electric potential due to each negative charge is negative. The basic equation to use is $V = k_e q / r$.

2. As in mechanics, only changes in potential energy are significant; hence, the point where you choose the potential energy to be zero is arbitrary. However, it is common practice to set $V = 0$ at infinity when this choice suits the particular situation.

A PROBLEM–SOLVING STRATEGY FOR CAPACITORS

1. Be careful with your choice of units. To calculate the capacitance of a device in farads, make sure that distances are in meters and use the SI value of ϵ_0.

2. **When two or more capacitors are connected in series**, they carry the same charge, but the potential differences across them are not the same unless they have equal values of capacitance. Their capacitances add as reciprocals, and the equivalent capacitance of the combination is always less than the smallest individual capacitor.

3. **When two or more capacitors are connected in parallel**, the potential differences across each are the same. The charge on each capacitor is proportional to its capacitance; hence, the capacitances add directly to give the equivalent capacitance of the parallel combination.

4. A complicated circuit consisting of capacitors can often be reduced to a simple circuit containing only one equivalent capacitor. To do so, examine the initial circuit and replace any capacitors in series or any in parallel using the rules of Steps 2 and 3 above. Draw a sketch of the new circuit after these changes have been made. Examine this new circuit and replace any series or parallel combinations. Continue this process until a single, equivalent capacitor is found.

5. If the charge on, or the electric potential difference across, one of the capacitors in a complicated circuit is to be found, start with the final circuit found in Step 4 and gradually work your way back through the circuits using $C = Q / \Delta V$ and the rules given in Steps 2 and 3 above. Make use of the successive sketches prepared in step 4 as the original circuit was simplified.

REVIEW CHECKLIST

▷ Understand that each point in the vicinity of a charge distribution can be characterized by a scalar quantity called the electric potential, V; and define the quantity, electrical potential difference.

▷ Calculate the electric potential difference between any two points in a uniform electric field and calculate the electric potential difference between any two points in the vicinity of a group of point charges.

▷ Calculate the electric potential energy associated with a group of point charges and define the unit of energy, the electron volt.

▷ Justify the claims that (i) all points on the surface and within a charged conductor are at the same potential and (ii) the electric field within a charged conductor is zero.

▷ Define the quantity, capacitance; evaluate the capacitance of a parallel plate capacitor of given area and plate separation.

▷ Determine the equivalent capacitance of a network of capacitors in series-parallel combination and calculate the final charge on each capacitor and the potential difference across each when a known potential is applied across the combination.

SOLUTIONS TO SELECTED END-OF-CHAPTER PROBLEMS

3. A potential difference of 90 mV exists between the inner and outer surfaces of a cell membrane. The inner surface is negative relative to the outer surface. How much work is required to eject a positive sodium ion (Na^+) from the interior of the cell?

Solution

Since the inner surface of the membrane is negative relative to the outer surface, the electric field within the membrane is directed from the outer surface toward the inner surface (recall that field lines originate on positive charges and terminate on negative charges). Thus, when the positive ion moves from the inner surface to the outer surface as it leaves the cell, it is being moved against the direction of the force exerted on it by the field. This means that its electrical potential energy is increasing, so the change in potential it undergoes is

$$\Delta V = +90 \text{ mV}$$

The work an external agent must do on the positive ion to overcome the influence of the electric field and move the ion out of the cell is given by

$$W = +q(\Delta V) = +(+e)(\Delta V) = +(1.60 \times 10^{-19} \text{ C})(90 \times 10^{-3} \text{ V})$$

or $\qquad W = 1.4 \times 10^{-20} \text{ J}$ ◊

7.   Oppositely charged parallel plates are separated by 5.33 mm. A potential difference of 600 V exists between the plates. (a) What is the magnitude of the electric field between the plates? (b) What is the magnitude of the force on an electron between the plates? (c) How much work must be done on the electron to move it to the negative plate if it is initially positioned 2.90 mm from the positive plate?

Solution

(a)   A uniform electric field, directed perpendicular to the plates, exists in the region between the plates and exerts a constant force $F = qE$ on a charged particle in this region. When the particle moves a distance d parallel to the field, going from one plate to the other, the work done by the field is $W = Fd\cos 0° = qEd$. Thus, the magnitude of the change in the potential energy of the particle is $\left|\Delta PE\right| = W = qEd$, and the magnitude of the potential difference between the plates is

$$\left|\Delta V\right| = \frac{\left|\Delta PE\right|}{q} = \frac{qEd}{q} = Ed$$

For the given set of plates, $\left|\Delta V\right| = 600$ V and $d = 5.33 \times 10^{-3}$ m. Therefore, the magnitude of the electric field between the plates is

$$E = \frac{\left|\Delta V\right|}{d} = \frac{600 \text{ V}}{5.33 \times 10^{-3} \text{ m}} = 1.13 \times 10^5 \text{ V/m} = 1.13 \times 10^5 \text{ N/C} \qquad \Diamond$$

(b)   The magnitude of the force on an electron located between the plates is

$$F = \left|q\right|E = eE = \left(1.60 \times 10^{-19} \text{ C}\right)\left(1.13 \times 10^5 \text{ N/C}\right) = 1.80 \times 10^{-14} \text{ N} \qquad \Diamond$$

(c)   The displacement of the electron as it moves from the initial position to the negative plate is

$$s = d - 2.90 \text{ mm} = 5.33 \text{ mm} - 2.90 \text{ mm} = 2.43 \text{ mm}$$

An applied force with magnitude $F = eE = 1.80 \times 10^{-14}$ N, directed toward the negative plate, must be used to offset the influence of the field and move the electron without acceleration. The work done by the applied force as the electron is moved to the negative plate is

$$W = Fs\cos 0° = \left(1.80 \times 10^{-14} \text{ N}\right)\left(2.43 \times 10^{-3} \text{ m}\right)\left(1.00\right) = 4.38 \times 10^{-17} \text{ J} \qquad \Diamond$$

13.   (a) Find the electric potential, taking zero at infinity, at the upper right corner (the corner without a charge) of the rectangle in Figure P16.13. (b) Repeat if the 2.00-μC charge is replaced with a charge of $-2.00\ \mu$C.

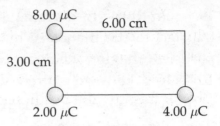

Figure P16.13

Solution

When the zero point of electric potential is at infinity, the potential at an observation point due to the presence of a collection of nearby point charges is

$$V = V_1 + V_2 + V_3 + \ldots = \Sigma V_i = \Sigma k_e \frac{q_i}{r_i}$$

where r_i is the distance from the position of the charge q_i to the observation point.

(a)   For the charge configuration shown in Figure P16.13, the electric potential at the upper right corner of the rectangle is

$$V = k_e \left(\frac{q_1}{r_1} + \frac{q_2}{r_2} + \frac{q_3}{r_3} \right)$$

where $q_1 = 8.00\ \mu$C, $r_1 = 6.00$ cm, $q_2 = 4.00\ \mu$C, $r_2 = 3.00$ cm,

$q_3 = 2.00\ \mu$C,   and $r_3 = \sqrt{r_1^2 + r_2^2} = 6.71$ cm

$$V = \left(8.99 \times 10^9\ \frac{\text{N} \cdot \text{m}^2}{\text{C}^2} \right) \left(\frac{8.00\ \mu\text{C}}{6.00\ \text{cm}} + \frac{4.00\ \mu\text{C}}{3.00\ \text{cm}} + \frac{2.00\ \mu\text{C}}{6.71\ \text{cm}} \right) \left(\frac{10^{-6}\ \text{C}}{1\ \mu\text{C}} \right) \left(\frac{1\ \text{cm}}{10^{-2}\ \text{m}} \right)$$

which gives $V = 2.67 \times 10^6$ V   ◊

(b)   If the charge $q_3 = 2.00\ \mu$C is replaced by a charge $q_3' = -2.00\ \mu$C, the new potential at the upper right corner is

$$V = \left(8.99 \times 10^9\ \frac{\text{N} \cdot \text{m}^2}{\text{C}^2} \right) \left(\frac{8.00\ \mu\text{C}}{6.00\ \text{cm}} + \frac{4.00\ \mu\text{C}}{3.00\ \text{cm}} - \frac{2.00\ \mu\text{C}}{6.71\ \text{cm}} \right) \left(\frac{10^{-6}\ \text{C}}{1\ \mu\text{C}} \right) \left(\frac{1\ \text{cm}}{10^{-2}\ \text{m}} \right)$$

or $V = 2.13 \times 10^6$ V   ◊

19. In Rutherford's famous scattering experiments that led to the planetary model of the atom, alpha particles (having charges of $+2e$ and masses of 6.64×10^{-27} kg) were fired toward a gold nucleus with charge $+79e$. An alpha particle, initially very far from the gold nucleus, is fired at 2.00×10^7 m/s directly toward the nucleus, as in Figure P16.19. How close does the alpha particle get to the gold nucleus before turning around? Assume the gold nucleus remains stationary.

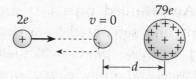

Figure P16.19

Solution

As the alpha particle approaches the gold nucleus, the only force acting on it is the electrical force exerted by the nucleus. This force is not constant, so it is difficult to use Newton's second law here. However, this is a conservative force, and the total mechanical energy of the system remains constant $\left(KE_f + PE_f = KE_i + PE_i\right)$.

Since the gold nucleus is "fixed" (that is, so massive in comparison to the alpha particle that its recoil may be ignored), the total kinetic energy is that of the alpha particle, $KE = \frac{1}{2}m_\alpha v_\alpha^2$. The potential energy of this pair of charges is $PE = k_e Q_1 Q_2 / r$, and the conservation of energy equation becomes

$$\frac{1}{2}m_\alpha v_f^2 + \frac{k_e Q_1 Q_2}{r_f} = \frac{1}{2}m_\alpha v_i^2 + \frac{k_e Q_1 Q_2}{r_i}$$

When the alpha particle is the minimum distance from the gold nucleus,

$$r_f = d \qquad \text{and} \qquad v_f = 0$$

Also, the alpha particle is initially "very far" from the nucleus,

so $$r_i \approx \infty \qquad \text{and} \qquad v_i = 2.00\times10^7 \text{ m/s}$$

The energy equation is then $$0 + \frac{k_e Q_1 Q_2}{d} = \frac{1}{2}m_\alpha v_i^2 + 0$$

and the closest the alpha particle gets to the nucleus is $d = 2k_e Q_1 Q_2 / m_\alpha v_i^2$:

$$d = \frac{2k_e (2e)(79e)}{m_\alpha v_i^2} = \frac{316\left(8.99\times10^9 \text{ N}\cdot\text{m}^2/\text{C}^2\right)\left(1.60\times10^{-19} \text{ C}\right)^2}{\left(6.64\times10^{-27} \text{ kg}\right)\left(2.00\times10^7 \text{ m/s}\right)^2} = 2.74\times10^{-14} \text{ m}$$ ◊

25. An air-filled capacitor consists of two parallel plates, each with an area of 7.60 cm^2 and separated by a distance of 1.80 mm. If a 20.0-V potential difference is applied to these plates, calculate (a) the electric field between the plates, (b) the capacitance, and (c) the charge on each plate.

Solution

The electric field in the region between a pair of parallel plates is uniform in magnitude and direction. Hence, the force exerted by this field on a particle of charge q located in this region is constant with magnitude $F = qE$. The potential difference between the plates (work per unit charge required to move charge from one plate to the other) is therefore given by

$$\Delta V = \frac{W}{q} = \frac{Fd}{q} = \frac{(qE)d}{q} = Ed$$

where d is the distance separating the plates.

(a) The electric field between the specified set of plates is

$$E = \frac{\Delta V}{d} = \frac{20.0 \text{ V}}{1.80 \times 10^{-3} \text{ m}} = 1.11 \times 10^4 \text{ V/m} = 11.1 \text{ kV/m}$$ ◊

The field is directed from the positive plate toward the negative plate. ◊

(b) The capacitance of an air-filled parallel-plate capacitor is $C = \kappa_{air} \epsilon_0 A/d$, where $\kappa_{air} \approx 1.00$, ϵ_0 is the permittivity of free space, and A is the surface area of one of the plates. For the given set of plates,

$$C = \frac{(1.00)(8.85 \times 10^{-12} \text{ C}^2/\text{N} \cdot \text{m}^2)(7.60 \text{ cm}^2)}{1.80 \times 10^{-3} \text{ m}} \left(\frac{1 \text{ m}}{10^2 \text{ cm}}\right)^2$$

or $C = 3.74 \times 10^{-12}$ F $= 3.74$ pF ◊

(c) The magnitude of the charge on each plate when the potential difference is $\Delta V = 20.0$ V is given by

$$Q = C(\Delta V) = (3.74 \times 10^{-12} \text{ F})(20.0 \text{ V}) = 7.47 \times 10^{-11} \text{ C} = 74.7 \text{ pC}$$ ◊

32. Two capacitors give an equivalent capacitance of 9.00 pF when connected in parallel and an equivalent capacitance of 2.00 pF when connected in series. What is the capacitance of each capacitor?

Solution

When two capacitors are connected in parallel, the equivalent capacitance of the combination is $C_{eq} = C_1 + C_2$. Therefore, we know the sum of the capacitances of our unknown capacitors:

$$C_1 + C_2 = 9.00 \text{ pF and } C_2 = 9.00 \text{ pF} - C_1 \qquad \text{[1]}$$

If two capacitors are connected in series, the equivalent capacitance is given by

$$\frac{1}{C_{eq}} = \frac{1}{C_1} + \frac{1}{C_2} = \frac{C_1 + C_2}{C_1 C_2} \text{ or } C_{eq} = \frac{C_1 C_2}{C_1 + C_2}$$

Thus, we may write a second equation: $\qquad \dfrac{C_1 C_2}{C_1 + C_2} = 2.00 \text{ pF} \qquad \text{[2]}$

Substituting Equation [1] into Equation [2] gives

$$\frac{C_1 (9.00 \text{ pF} - C_1)}{9.00 \text{ pF}} = 2.00 \text{ pF}$$

simplifying to $\qquad C_1^2 - (9.00 \text{ pF})C_1 + 18.0 \ (\text{pF})^2 = 0$

or $\qquad (C - 3.00 \text{ pF})(C - 6.00 \text{ pF}) = 0$

The solutions are $\qquad C_1 = 3.00 \text{ pF} \qquad \text{and} \qquad C_2 = 6.00 \text{ pF} \qquad \Diamond$

or $\qquad C_1 = 6.00 \text{ pF} \qquad \text{and} \qquad C_2 = 3.00 \text{ pF} \qquad \Diamond$

35. Find the charge on each of the capacitors in Figure P16.35.

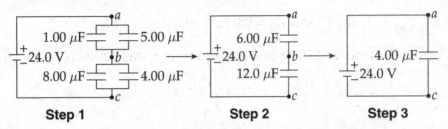

Figure P16.35

Solution

First, we reduce the circuit, in several steps, to a single equivalent capacitor connected to the battery by applying the rules for combining capacitors connected in series and in parallel as shown below:

Step 1　**Step 2**　**Step 3**

Using the figure of Step 3, we find the total charge stored between points a and c to be:

$$Q_{ac} = C_{ac}\Delta V_{ac} = (4.00\ \mu\text{F})(24.0\ \text{V}) = 96.0\ \mu\text{C}$$

Since capacitors connected in series (as in the figure of Step 2) have equal charges, with the charge on each being equal to the total charge stored in the combination, we know that $Q_{ab} = Q_{bc} = Q_{ac} = 96.0\ \mu\text{F}$. Thus,

$$\Delta V_{ab} = \frac{Q_{ab}}{C_{ab}} = \frac{96.0\ \mu\text{C}}{6.00\ \mu\text{F}} = 16.0\ \text{V} \quad\text{and}\quad \Delta V_{bc} = \frac{Q_{bc}}{C_{bc}} = \frac{96.0\ \mu\text{C}}{12.0\ \mu\text{F}} = 8.00\ \text{V}$$

Then, using the figure of Step 1, we find the charge on each of the four original capacitors:

For the 1.00-μF: $\quad Q_1 = C_1\Delta V_{ab} = (1.00\ \mu\text{F})(16.0\ \text{V}) = 16.0\ \mu\text{C}$ ◊

For the 5.00-μF: $\quad Q_5 = C_5\Delta V_{ab} = (5.00\ \mu\text{F})(16.0\ \text{V}) = 80.0\ \mu\text{C}$ ◊

For the 8.00-μF: $\quad Q_8 = C_8\Delta V_{bc} = (8.00\ \mu\text{F})(8.00\ \text{V}) = 64.0\ \mu\text{C}$ ◊

For the 4.00-μF: $\quad Q_4 = C_4\Delta V_{bc} = (4.00\ \mu\text{F})(8.00\ \text{V}) = 32.0\ \mu\text{C}$ ◊

Note that $Q_1 + Q_5 = Q_8 + Q_4 = 96\ \mu\text{C} = Q_{ac}$ as one should expect.

39. A 1.00-μF capacitor is charged by being connected across a 10.0-V battery. It is then disconnected from the battery and connected across an uncharged 2.00-μF capacitor. Determine the resulting charge on each capacitor.

Solution

The charge stored on the 1.00-μF capacitor by the battery is

$$Q = C_1 \Delta V_{battery} = (1.00 \ \mu F)(10.0 \ V) = 10.0 \ \mu C$$

If this capacitor is now carefully disconnected from the battery and then connected in parallel with an uncharged 2.00-μF, the charge Q remains stored as the total charge in the parallel combination. However, a portion of it will now be stored on the 2.00-μF capacitor with the remainder left on the 1.00-μF capacitor.

The equivalent capacitance of the parallel combination is

$$C_{eq} = C_1 + C_2 = 3.00 \ \mu F$$

and with a total stored charge of $Q = 10.0 \ \mu C$, the potential difference across each capacitor in the combination is

$$\Delta V_p = \frac{Q}{C_{eq}} = \frac{10.0 \ \mu C}{3.00 \ \mu F} = 3.33 \ V$$

The charged stored on each capacitor in the parallel combination is then

For the 1.00-μF: $\qquad Q_1 = C_1 \Delta V_p = (1.00 \ \mu F)(3.33 \ V) = 3.33 \ \mu C$ $\qquad \Diamond$

For the 2.00-μF: $\qquad Q_2 = C_2 \Delta V_p = (2.00 \ \mu F)(3.33 \ V) = 6.67 \ \mu C$ $\qquad \Diamond$

Note that $Q_1 + Q_2 = Q = 10.0 \ \mu C$ as it must in order to conserve charge.

45. Consider the parallel-plate capacitor formed by the Earth and a cloud layer as described in Problem 16.23. Assume this capacitor will discharge (i.e., produce lightning) when the electric field strength between the plates reaches 3.0×10^6 N/C. What is the energy released if the capacitor discharges completely during a lighting strike?

Solution

We model the Earth and cloud layer as a parallel-plate capacitor with the plates separated by $d = 800$ m, and each plate having surface area $A = 1.0 \times 10^6$ m^2. The capacitance of this capacitor is

$$C = \frac{\kappa_{air}\,\epsilon_0\,A}{d} = \frac{(1.00)(8.85 \times 10^{-12}\ \text{C}^2/\text{N} \cdot \text{m}^2)(1.0 \times 10^6\ \text{m}^2)}{800\ \text{m}} = 1.1 \times 10^{-8}\ \text{F}$$

A lightning strike will occur and discharge the capacitor when the electric field between the plates reaches a magnitude of $E = 3.0 \times 10^6$ N/C $= 3.0 \times 10^6$ V/m. The potential difference between the Earth and cloud at that instant is

$$\Delta V = E d = (3.0 \times 10^6\ \text{V/m})(800\ \text{m}) = 2.4 \times 10^9\ \text{V}$$

Thus, the energy stored in the capacitor and available for release during the lightning strike is

$$W = \tfrac{1}{2}C(\Delta V)^2 = \tfrac{1}{2}(1.1 \times 10^{-8}\ \text{F})(2.4 \times 10^9\ \text{V})^2 = 3.2 \times 10^{10}\ \text{J} \qquad \Diamond$$

49.   Determine (a) the capacitance and (b) the maximum voltage that can be applied to a Teflon®-filled parallel-plate capacitor having a plate area of 175 cm^2 and an insulation thickness of $0.040\,0 \text{ mm}$.

Solution

(a)   A parallel-plate capacitor with a dielectric material between the plates has a capacitance of $C = \kappa \in_0 A/d$ where κ is the dielectric constant of the material, $\in_0$ is the permittivity of free space, A is the area of one of the plates, and d is the distance separating the plates. From Table 16.1 in the textbook, the dielectric constant of Teflon® is $\kappa = 2.1$, so the capacitance of this parallel-plate capacitor is

$$C = \frac{(2.1)\left(8.85 \times 10^{-12} \text{ C}^2/\text{N} \cdot \text{m}^2\right)\left(175 \times 10^{-4} \text{ m}^2\right)}{0.040\,0 \times 10^{-3} \text{ m}} = 8.1 \times 10^{-9} \text{ F} = 8.1 \text{ nF} \qquad \lozenge$$

(b)   The dielectric strength of a material is the maximum electric field that material can withstand before it breaks down and begins to conduct. From Table 16.1, the dielectric strength for Teflon® is $E_{max} = 60 \times 10^6 \text{ V/m}$. The potential difference between the plates of a parallel-plate capacitor is $\Delta V = Ed$ where E is the electric field strength, and d is the distance between the plates. Thus, the maximum voltage that can be applied to this Teflon®-filled capacitor is

$$\Delta V_{max} = E_{max}d = \left(60 \times 10^6 \text{ V/m}\right)\left(0.040\,0 \times 10^{-3} \text{ m}\right)$$

or $\Delta V_{max} = 2.4 \times 10^3 \text{ V} = 2.4 \text{ kV}$ $\qquad \lozenge$

58. A spherical capacitor consists of a spherical conducting shell of radius b and charge $-Q$ concentric with a smaller conducting sphere of radius a and charge Q. (a) Find the capacitance of this device. (b) Show that as the radius b of the outer sphere approaches infinity, the capacitance approaches the value $a/k_e = 4\pi \epsilon_0\, a$.

Solution

(a) Due to spherical symmetry, the charge on each of the concentric spherical shells will be uniformly distributed over that shell. Inside a spherical surface having a uniform charge distribution, the electric field due to the charge on that surface is zero. Thus, in this region, the potential due to the charge on that surface is constant and equal to the potential at the surface. Outside a spherical surface having a uniform charge distribution, the potential due to the charge on that surface is given by $V = k_e q/r$, where r is the distance from the center of that surface, and q is the charge on that surface.

In the region **between** a pair of concentric spherical shells, with the inner shell having charge $+Q$ and the outer shell having radius b and charge $-Q$, the total electric potential is given by

$$V = V_{\substack{due\ to \\ inner\ shell}} + V_{\substack{due\ to \\ outer\ shell}} = \frac{k_e Q}{r} + \frac{k_e(-Q)}{b} = k_e Q\left(\frac{1}{r} - \frac{1}{b}\right)$$

The potential difference between the two shells is therefore,

$$\Delta V = V\big|_{r=a} - V\big|_{r=b} = k_e Q\left(\frac{1}{a} - \frac{1}{b}\right) - k_e Q\left(\frac{1}{b} - \frac{1}{b}\right) = k_e Q\left(\frac{b-a}{ab}\right)$$

The capacitance of this device is given by

$$C = \frac{Q}{\Delta V} = \frac{ab}{k_e(b-a)} \qquad\qquad \lozenge$$

(b) When b is very large in comparison to a, we have $b - a \approx b$. Thus, in the limit as $b \to \infty$, the capacitance found above becomes

$$C \to \frac{ab}{k_e(b)} = \frac{a}{k_e} = 4\pi \epsilon_0\, a \qquad\qquad \lozenge$$

65.   Consider a parallel-plate capacitor with charge Q and area A, filled with dielectric material having dielectric constant κ. It can be shown that the magnitude of the attractive force exerted on each plate by the other is $F = Q^2/(2\kappa\,\epsilon_0\,A)$. When a potential difference of 100 V exists between the plates of an air-filled 20-μF parallel-plate capacitor, what force does each plate exert on the other if they are separated by 2.0 mm?

Solution

The capacitance of a parallel-plate capacitor is $C = \kappa\,\epsilon_0\,A/d$, where ϵ_0 is the permittivity of free space and d is the distance separating the plates. Thus, the product $\kappa\,\epsilon_0\,A$ may be written as $\kappa\,\epsilon_0\,A = Cd$, and the given force equation becomes

$$F = \frac{Q^2}{2\kappa\,\epsilon_0\,A} = \frac{Q^2}{2Cd}$$

The magnitude of the charge Q on either plate of the capacitor, when a potential difference ΔV exists between the plates, is given by $Q = C(\Delta V)$. Therefore, the force equation may be rewritten as

$$F = \frac{Q^2}{2Cd} = \frac{C^2(\Delta V)^2}{2Cd} = \frac{C(\Delta V)^2}{2d}$$

If a parallel-plate capacitor having plate separation $d = 2.0$ mm $= 2.0\times10^{-3}$ m and capacitance $C = 20\ \mu$F $= 20\times10^{-6}$ F has a potential difference of $\Delta V = 100$ V maintained between its plates, the force with which one plate attracts the other is

$$F = \frac{C(\Delta V)^2}{2d} = \frac{(20\times10^{-6}\ \text{F})(100\ \text{V})^2}{2(2.0\times10^{-3}\ \text{m})} = 50\ \text{N}$$ ◊

Chapter 17

Current and Resistance

NOTES FROM SELECTED CHAPTER SECTIONS

17.1 Electric Current

The direction of conventional current is designated as the direction in which positive charges would flow. *In an ordinary metal conductor, the direction of conventional current will be opposite the direction of flow of electrons (which are the charge carriers in this case).*

17.2 A Microscopic View: Current and Drift Speed

In the classical model of electronic conduction in a metal, electrons are treated like molecules in a gas. In the absence of an electric field, electrons have an average velocity of zero.

Under the influence of an electric field, the electrons move along a direction opposite the direction of the applied field with a drift velocity, v_d. The drift velocity of electrons in a conductor is proportional to the current and inversely proportional to the number of free electrons per unit volume, n. *The drift speed is much smaller than the average speed of electrons between collisions.*

17.4 Resistance and Ohm's Law

When a potential difference, ΔV, is applied across the ends of a metallic conductor, the current in the conductor is found to be proportional to the applied voltage. If the proportionality is exact, we can write $\Delta V = IR$ where the proportionality constant, R, is called the resistance of the conductor. In fact, we define this resistance as the ratio of the voltage across the conductor to the current it carries. For many materials, including most metals, experiments show that the resistance is constant over a wide range of applied voltages. This statement is known as Ohm's law.

Ohm's law is not a fundamental law of nature but an empirical relationship that is valid only for certain materials. Materials that obey Ohm's law, and hence have a constant resistance over a wide range of voltages, are said to be ohmic. Materials that do not obey Ohm's law are nonohmic.

Resistance has the SI units volts per ampere, called ohms (Ω). Thus, if a potential difference of 1 V across a conductor produces a current of 1 A, the resistance of the conductor is 1 Ω.

17.5   Resistivity

A current can be established in a conductor by applying a potential difference between two points in the conductor (for example between the ends of a wire). The average velocity of electrons moving through the conductor is limited by collisions with atoms of the conducting material. The overall effect of the conductor inhibiting electron flow is due to the resistivity of the conducting material, ρ, which depends on the molecular structure and temperature of the conductor. *Resistivity, ρ, is characteristic of a particular **type** of material (for example copper), and the value of ρ for a material does not depend on the size or shape of the material.* Electrical resistance, R, is associated with a particular sample of material (for example a cylinder of copper of specific length and cross-sectional area). Good conductors have low values of ρ.

17.6   Temperature Variation of Resistance

The rate at which resistivity of a material (and resistance of a conductor) changes with temperature depends on a parameter called the temperature coefficient of resistance, α. For most metals, α is positive and the resistivity changes approximately linearly with temperature over a limited temperature range; α has a negative value for most semiconductors.

EQUATIONS AND CONCEPTS

Electric current is defined as the rate at which charge flows through a cross section of a conductor. *Under the influence of an electric field, electric charges will move through conducting gases, liquids, and solids.* The SI unit of current is the ampere (A).

$$I = \frac{\Delta Q}{\Delta t} \qquad (17.1)$$

$$1 \text{ A} = 1 \text{ C/s}$$

The **drift velocity**, v_d (velocity of the charge carriers in a conductor) is actually an average value of the individual velocities.

$$I = nqv_dA \qquad (17.2)$$

n = the number of mobile charge carriers per unit volume of conductor

In **Ohm's law** as expressed in Equation 17.3, R is understood to be independent of ΔV. This form of Ohm's law is valid only for ohmic materials which have a linear current-voltage relationship over a wide range of applied voltages.

$$R = \frac{\Delta V}{I} \qquad (17.3)$$

The **resistance, R of a given conductor** made of a homogeneous material of uniform cross section, A, can be expressed in terms of the dimensions of the conductor and an intrinsic property of the material of which the conductor is made called its **resistivity**. *The value of the resistivity of a given material depends on the electronic structure of the material and on the temperature.*

$$R = \rho\frac{\ell}{A} \qquad (17.5)$$

ρ = resistivity

The **ohm** is the SI unit of resistance. If a potential difference of 1 V across a conductor produces a current of 1 A, the resistance of the conductor is 1 ohm. Resistivity is expressed in SI units of ohm-meters ($\Omega \cdot m$).

$$1\,\Omega = 1\,V/A$$

The **resistivity**, and therefore the resistance, of a conductor varies with temperature. Over a limited range of temperatures, this variation is approximately linear. *Semiconductors (for example, carbon) are characterized by a negative temperature coefficient of resistivity, and in these materials the resistance decreases as the temperature increases.*

$$\rho = \rho_0[1 + \alpha(T - T_0)] \qquad (17.6)$$

$$R = R_0[1 + \alpha(T - T_0)] \qquad (17.7)$$

α = temperature coefficient of resistivity

T_0 is a reference temperature (usually taken to be 20.0 °C)

Joule's law can be used to calculate the power delivered to a resistor or other current-carrying device. In an ohmic resistor, the power can be expressed in alternative forms.

$$\mathcal{P} = I\Delta V \tag{17.8}$$

$$\mathcal{P} = I^2 R = \frac{(\Delta V)^2}{R} \tag{17.9}$$

The **SI unit of power** is the watt (W).

$$1\text{ W} = 1\text{ J/s} = 1\text{ V}\cdot\text{A}$$

The **kilowatt-hour** is the quantity of energy consumed in one hour at a constant use rate (or power) of 1 kW. It is the unit used by electric companies to bill for energy consumption.

$$1\text{ kWh} = 3.60 \times 10^6\text{ J} \tag{17.10}$$

REVIEW CHECKLIST

▷ Define the term "electric current" in terms of rate of charge flow and its corresponding unit of measure, the ampere. Calculate electron drift velocity in a conductor of specified characteristics carrying a current, I.

▷ Determine the resistance of a conductor using Ohm's law. Also, calculate the resistance based on the physical characteristics of a conductor. Distinguish between ohmic and nonohmic conductors.

▷ Make calculations of the variation of resistance with temperature, which involves the concept of the temperature coefficient of resistivity.

▷ Sketch a simple single-loop circuit to illustrate the use of basic circuit element symbols and direction of conventional current.

▷ Use Joule's law to calculate the power dissipated in a resistor.

SOLUTIONS TO SELECTED END-OF-CHAPTER PROBLEMS

1. If a current of 80.0 mA exists in a metal wire, how many electrons flow past a given cross section of the wire in 10.0 min? Sketch the direction of the current and the direction of the electrons' motion.

Solution

The current in a conductor is defined as $I = \dfrac{\Delta Q}{\Delta t}$

where ΔQ is the amount of charge passing a fixed cross section of the conductor in time Δt. Thus, if $I = 80.0 \text{ mA} = 80.0 \times 10^{-3} \text{ A} = 80.0 \times 10^{-3} \text{ C/s}$, the magnitude of the charge passing a fixed cross section of the wire in time $\Delta t = 10.0 \text{ min}$ is

$$|\Delta Q| = I(\Delta t) = \left(80.0 \times 10^{-3} \text{ C/s}\right)(10.0 \text{ min})\left(\frac{60.0 \text{ s}}{1 \text{ min}}\right) = 48.0 \text{ C}$$

The number of electrons that has a total charge of this magnitude is

$$N = \frac{|\Delta Q|}{e} = \frac{48.0 \text{ C}}{1.60 \times 10^{-19} \text{ C/electron}} = 3.00 \times 10^{20} \text{ electrons}$$ ◊

The direction of conventional current is the direction that positive charges would flow through the conductor if they were free to move. That is, the current is in the direction of the electric field within the conductor. Free electrons, being negative charges, experience an electrical force and move in the direction opposite to that of the electric field, and hence opposite to the direction of the conventional current. The requested sketch, showing the relative directions of the current and the electrons' motion, is given below:

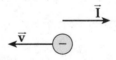

7.   A 200-km-long high-voltage transmission line 2.0 cm in diameter carries a steady current of 1 000 A. If the conductor is copper with a free charge density of 8.5×10^{28} electrons per cubic meter, how many years does it take one electron to travel the full length of the cable?

Solution

The drift speed of charge carriers in a conductor carrying current I is

$$v_d = \frac{I}{nqA}$$

where n is the number of mobile charge carriers per unit volume, q is the magnitude of the charge of each carrier, and A is the cross-sectional area of the conductor.

The transmission line has a diameter of 2.0 cm, so its radius is $r = 1.0 \text{ cm} = 1.0 \times 10^{-2}$ m, and the cross-sectional area is

$$A = \pi r^2 = \pi \left(1.0 \times 10^{-2} \text{ m}\right)^2 = \pi \times 10^{-4} \text{ m}^2$$

In metallic conductors, the mobile charge carriers are electrons with charge of magnitude $q = e = 1.60 \times 10^{-19}$ C, and the density of free electrons in this copper material is $n = 8.5 \times 10^{28}$ per cubic meter. Thus, when this cable carries a current $I = 1\,000$ A, the drift speed of the electrons is

$$v_d = \frac{1\,000 \text{ C/s}}{\left(8.5 \times 10^{28} \text{ m}^{-3}\right)\left(1.60 \times 10^{-19} \text{ C}\right)\left(\pi \times 10^{-4} \text{ m}^2\right)} = 2.3 \times 10^{-4} \text{ m/s}$$

and the time the electrons require to drift the full 200-km length of the cable is

$$\Delta t = \frac{L}{v_d} = \frac{200 \times 10^3 \text{ m}}{2.3 \times 10^{-4} \text{ m/s}} = \left(8.5 \times 10^8 \text{ s}\right)\left(\frac{1 \text{ yr}}{3.156 \times 10^7 \text{ s}}\right) = 27 \text{ yr} \qquad \Diamond$$

17.   A wire 50.0 m long and 2.00 mm in diameter is connected to a source with a potential difference of 9.11 V, and the current is found to be 36.0 A. Assume a temperature of 20°C, and, using Table 17.1, identify the metal out of which the wire is made.

Solution

From Ohm's law, the resistance of the wire must be

$$R = \frac{\Delta V}{I} = \frac{9.11 \text{ V}}{36.0 \text{ A}} = 0.253 \ \Omega$$

Then, from $R = \rho L/A$, the resistivity of the material making up the wire is found to be

$$\rho = \frac{RA}{L} = \frac{R\left(\pi d^2/4\right)}{L} = \frac{\left(0.253 \ \Omega\right)\pi\left(2.00\times10^{-3} \text{ m}\right)^2}{4\left(50.0 \text{ m}\right)} = 1.59\times10^{-8} \ \Omega\cdot\text{m}$$

Comparing this result with the resistivities of various materials found in Table 17.1 of the textbook, we conclude that the metal contained in the wire is silver.   ◊

21.   While taking photographs in Death Valley on a day when the temperature is 58.0 °C, Bill Hiker finds that a certain voltage applied to a copper wire produces a current of 1.000 A. Bill then travels to Antarctica and applies the same voltage to the same wire. What current does he register there if the temperature is –88.0 °C? Assume that no change occurs in the wire's shape and size.

Solution

The same applied voltage, ΔV, is used in the initial test (in Death Valley) and in the final test (conducted in Antarctica). Therefore, from Ohm's law,

$$\Delta V = I_{DV}R_{DV} = I_A R_A \quad \text{and} \quad I_A = \left(\frac{R_{DV}}{R_A}\right)I_{DV} \tag{1}$$

At temperature T, the resistance of a conducting element is given by

$$R = R_0\left[1 + \alpha\left(T - T_0\right)\right]$$

where α is the temperature coefficient of resistivity for the material in use, and R_0 is the resistance of that element at the reference temperature (normally 20.0 °C).

For copper, $\alpha = 3.90 \times 10^{-3}\ (°C)^{-1}$, so Equation [1] gives the current in Antarctica as

$$I_A = \left(\frac{1 + \alpha\left(T_{DV} - T_0\right)}{1 + \alpha\left(T_A - T_0\right)}\right)I_{DV}$$

or $I_A = \left(\dfrac{1 + \left[3.90 \times 10^{-3}\ (°C)^{-1}\right]\left(58.0\ °C - 20.0\ °C\right)}{1 + \left[3.90 \times 10^{-3}\ (°C)^{-1}\right]\left(-88.0\ °C - 20.0\ °C\right)}\right)(1.000\ A) = 1.98\ A$ ◊

27. (a) A 34.5-m length of copper wire at 20.0 °C has a radius of 0.25 mm. If a potential difference of 9.0 V is applied across the length of the wire, determine the current in the wire. (b) If the wire is heated to 30.0 °C while the 9.0-V potential difference is maintained, what is the resulting current in the wire?

Solution

(a) For copper, the resistivity is $\rho = 1.7 \times 10^{-8} \ \Omega \cdot m$

and the temperature coefficient of resistivity is $\alpha = 3.9 \times 10^{-3} \ (°C)^{-1}$

Therefore, the resistance of this wire at $T = T_0 = 20.0$ °C is

$$R_0 = \frac{\rho L}{A} = \frac{\rho L}{\pi r^2} = \frac{(1.7 \times 10^{-8} \ \Omega \cdot m)(34.5 \ m)}{\pi (0.25 \times 10^{-3} \ m)^2} = 3.0 \ \Omega$$

When a 9.0-V potential difference is applied across the length of this wire, Ohm's law gives the resulting current as

$$I_0 = \frac{\Delta V}{R_0} = \frac{9.0 \ V}{3.0 \ \Omega} = 3.0 \ A \qquad \Diamond$$

(b) At $T = 30.0$ °C, the resistance is given by $R = R_0 \left[1 + \alpha (T - T_0) \right]$ and the current becomes

$$I = \frac{\Delta V}{R} = \frac{9.0 \ V}{(3.0 \ \Omega) \left[1 + \left(3.9 \times 10^{-3} \ (°C)^{-1} \right)(30.0 \ °C - 20.0 \ °C) \right]} = 2.9 \ A \qquad \Diamond$$

35.   The heating element of a coffeemaker operates at 120 V and carries a current of 2.00 A. Assuming that the water absorbs all of the energy converted by the resistor, calculate how long it takes to heat 0.500 kg of water from room temperature (23.0°C) to the boiling point.

Solution

The energy required to raise the temperature of the water from $T_i = 23.0°C$ to the boiling point $(T = 100°C)$ is

$$Q = mc(\Delta T) = (0.500 \text{ kg})(4\,186 \text{ J/kg} \cdot °C)(100°C - 23.0°C) = 1.61 \times 10^5 \text{ J}$$

The rate at which the heating element is converting electrical potential energy into internal energy of the water is

$$\mathcal{P} = (\Delta V)I = (120 \text{ V})(2.00 \text{ A}) = 240 \text{ W} = 240 \text{ J/s}$$

Thus, the time required to bring the water to a boil will be

$$\Delta t = \frac{Q}{\mathcal{P}} = \frac{1.61 \times 10^5 \text{ J}}{240 \text{ J/s}} = (672 \text{ s})\left(\frac{1 \text{ min}}{60.0 \text{ s}}\right) = 11.2 \text{ min}$$

◊

39.   A copper cable is designed to carry a current of 300 A with a power loss of 2.00 W/m. What is the required radius of this cable?

Solution

The rate at which electrical power is dissipated when current flows through a resistance is given by

$$\mathcal{P} = I^2 R$$

Thus, if the cable is to have a power loss per unit length of 2.00 W/m when carrying a current of $I = 300$ A, the resistance per unit length of the cable must be

$$\frac{R}{L} = \frac{\mathcal{P}/L}{I^2} = \frac{2.00 \text{ W/m}}{(300 \text{ A})^2} = 2.22 \times 10^{-5} \ \Omega/\text{m}$$

From $R = \rho \dfrac{L}{A} = \rho \dfrac{L}{\pi r^2}$, the required radius of the copper cable is

$$r = \sqrt{\frac{\rho}{\pi (R/L)}} = \sqrt{\frac{1.7 \times 10^{-8} \ \Omega \cdot \text{m}}{\pi (2.22 \times 10^{-5} \ \Omega/\text{m})}} = 1.6 \times 10^{-2} \text{ m} = 1.6 \text{ cm} \qquad \Diamond$$

45. An 11-W energy-efficient fluorescent lamp is designed to produce the same illumination as a conventional 40-W lamp. How much does the energy-efficient lamp save during 100 hours of use? Assume a cost of $0.080/kWh for electrical energy.

Solution

The energy used by a 40-watt lamp in 100 hours of use is

$$E_{40} = \mathcal{P}_{40}(\Delta t) = (0.040 \text{ kW})(100 \text{ h}) = 4.0 \text{ kWh}$$

and the energy used by the 11-W fluorescent lamp in the same time is

$$E_{11} = \mathcal{P}_{11}(\Delta t) = (0.011 \text{ kW})(100 \text{ h}) = 1.1 \text{ kWh}$$

so the energy saved by the fluorescent lamp over the conventional lamp is

$$E_{saved} = E_{40} - E_{11} = 4.0 \text{ kWh} - 1.1 \text{ kWh} = 2.9 \text{ kWh}$$

At a rate of $0.080/kWh for electrical energy, this represents a monetary savings of

$$Savings = E_{saved} \times rate = (2.9 \text{ kWh})\left(\frac{\$0.080}{\text{kWh}}\right) = \$0.23 = 23 \text{ cents} \qquad \Diamond$$

49. A particular wire has a resistivity of 3.0×10^{-8} $\Omega \cdot m$ and a cross-sectional area of 4.0×10^{-6} m^2. A length of this wire is to be used as a resistor that will develop 48 W of power when connected across a 20-V battery. What length of wire is required?

Solution

As the moving charges constituting a current I undergo a decrease in electric potential of ΔV, the power dissipated is $\mathcal{P} = (\Delta V) I$. For an ohmic resistance, we may use Ohm's law and write this as

$$\mathcal{P} = (\Delta V) I = (IR) I = I^2 R \quad \text{or} \quad \mathcal{P} = (\Delta V) I = (\Delta V)(\Delta V/R) = (\Delta V)^2 / R$$

The resistance of length L of a wire having resistivity ρ and cross-sectional area A is $R = \rho L / A$. Therefore, the last expression for the power dissipated becomes

$$\mathcal{P} = \frac{(\Delta V)^2}{R} = \frac{A (\Delta V)^2}{\rho L}$$

If the specified wire is to dissipate 48 W of power when connected to a 20-V battery, its length must be

$$L = \frac{A (\Delta V)^2}{\rho \mathcal{P}} = \frac{\left(4.0 \times 10^{-6} \ m^2\right)(20 \ V)^2}{\left(3.0 \times 10^{-8} \ W \cdot m\right)(48 \ W)} = 1.1 \times 10^3 \ m = 1.1 \ km \qquad \lozenge$$

55. An electric car is designed to run off a bank of 12.0-V batteries with a total energy storage of 2.00×10^7 J. (a) If the electric motor draws 8.00 kW, what is the current delivered to the motor? (b) If the electric motor draws 8.00 kW as the car moves at a steady speed of 20.0 m/s, how far will the car travel before it is "out of juice"?

Solution

(a)   When current I flows between points maintained at a potential difference ΔV, the power delivered is $\mathcal{P} = (\Delta V)I$. Thus, if the batteries maintain a potential difference of 12.0 V across the motor while that motor delivers 8.00 kW of power, the current must be

$$I = \frac{\mathcal{P}}{\Delta V} = \frac{8.00 \times 10^3 \text{ W}}{12.0 \text{ V}} = 667 \text{ A} \qquad \Diamond$$

(b)   The motor uses energy at the rate $\mathcal{P} = 8.00 \text{ kW} = 8.00 \times 10^3 \text{ J/s}$. At this rate, the time required to exhaust the energy stored in the batteries will be

$$t = \frac{E}{\mathcal{P}} = \frac{2.00 \times 10^7 \text{ J}}{8.00 \times 10^3 \text{ J/s}} = 2.5 \times 10^3 \text{ s}$$

During this time, the car can travel a distance of

$$d = vt = (20.0 \text{ m/s})(2.50 \times 10^3 \text{ s}) = 50.0 \times 10^3 \text{ m} = 50.0 \text{ km} \qquad \Diamond$$

61.   A resistor is constructed by forming a material of resistivity 3.5×10^5 $\Omega \cdot$m into the shape of a hollow cylinder of length 4.0 cm and inner and outer radii 0.50 cm and 1.2 cm, respectively. In use, a potential difference is applied between the ends of the cylinder, producing a current parallel to the length of the cylinder. Find the resistance of the cylinder.

Solution

The cross-sectional area of the material the current passes through (shaded in the sketch at the right) is given by

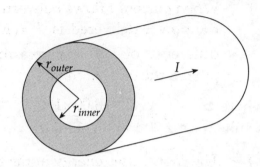

$$A = \pi r_{outer}^2 - \pi r_{inner}^2$$

$$= \pi \left[(1.2 \text{ cm})^2 - (0.50 \text{ cm})^2 \right]$$

or $$A = \pi \left[(1.2 \times 10^{-2} \text{ m})^2 - (0.50 \times 10^{-2} \text{ m})^2 \right]$$

This hollow cylindrical conductor has a length $L = 4.0$ cm $= 4.0 \times 10^{-2}$ m and is made from material having resistivity $\rho = 3.5 \times 10^5$ $\Omega \cdot$m. Therefore, $R = \rho L / A$ gives the resistance of the cylinder as

$$R = \frac{\left(3.5 \times 10^5 \ \Omega \cdot \text{m} \right) \left(4.0 \times 10^{-2} \text{ m} \right)}{\pi \left[\left(1.2 \times 10^{-2} \text{ m} \right)^2 - \left(0.50 \times 10^{-2} \text{ m} \right)^2 \right]} = 3.7 \times 10^7 \ \Omega = 37 \text{ M}\Omega \qquad \Diamond$$

66. When a straight wire is heated, its resistance changes according to the equation

$$R = R_0 \left[1 + \alpha \left(T - T_0 \right) \right]$$

(Eq. 17.7), where α is the temperature coefficient of resistivity. (a) Show that a more precise result, which includes the fact that the length and area of a wire change when it is heated, is

$$R = \frac{R_0 \left[1 + \alpha \left(T - T_0 \right) \right] \left[1 + \alpha' \left(T - T_0 \right) \right]}{1 + 2\alpha' \left(T - T_0 \right)}$$

where α' is the coefficient of linear expansion. (See Chapter 10.) (b) Compare the two results for a 2.00-m-long copper wire of radius 0.100 mm, starting at 20.0°C and heated to 100.0°C.

Solution

(a)   The resistance of a conductor is given by $R = \rho L / A$. The approximate expression $R = R_0 \left[1 + \alpha \left(T - T_0 \right) \right]$ was obtained by recognizing that the resistivity of the material changes with temperature according to $\rho = \rho_0 \left[1 + \alpha \left(T - T_0 \right) \right]$ where ρ_0 is the resistivity at temperature T_0. However, it was assumed that changes in the length L and the cross-sectional area A would be negligible.

To obtain a more precise expression for the temperature variation of resistance, we need to include the variation of L and A with changes in temperature. If L_0 and A_0 are the length and cross-sectional area at temperature T_0, we know from Chapter 10 that their values at temperature T will be $L = L_0 \left[1 + \alpha' \left(T - T_0 \right) \right]$ and $A = A_0 \left[1 + 2\alpha' \left(T - T_0 \right) \right]$ where α' is the coefficient of linear expansion of the conducting material.

Thus, the more precise expression for the resistance at temperature T is

$$R = \frac{\rho L}{A} = \frac{\rho_0 \left[1 + \alpha \left(T - T_0 \right) \right] L_0 \left[1 + \alpha' \left(T - T_0 \right) \right]}{A_0 \left[1 + 2\alpha' \left(T - T_0 \right) \right]}$$

$$= \left(\frac{\rho_0 L_0}{A_0} \right) \frac{\left[1 + \alpha \left(T - T_0 \right) \right] \left[1 + \alpha' \left(T - T_0 \right) \right]}{\left[1 + 2\alpha' \left(T - T_0 \right) \right]}$$

or
$$R = \frac{R_0 \left[1 + \alpha\left(T - T_0\right)\right]\left[1 + \alpha'\left(T - T_0\right)\right]}{\left[1 + 2\alpha'\left(T - T_0\right)\right]}$$
◊

where $R_0 = \rho_0 L_0 / A_0$ is the resistance at temperature T_0.

(b)   At $T_0 = 20.0°C$, the resistance of a copper wire of length $L_0 = 2.00$ m and radius $r_0 = 0.100$ mm is

$$R_0 = \frac{\left(1.70 \times 10^{-8}\ \Omega \cdot m\right)\left(2.00\ m\right)}{\pi \left(1.00 \times 10^{-4}\ m\right)^2} = 1.082\ \Omega$$

Then, at $T = 100°C$ the less accurate expression $R = R_0 \left[1 + \alpha\left(T - T_0\right)\right]$ gives:

$$R = \left(1.082\ \Omega\right)\left[1 + \left(3.90 \times 10^{-3}\ °C^{-1}\right)\left(80.0°C\right)\right] = 1.420\ \Omega$$
◊

while the more precise expression gives:

$$R = \frac{\left(1.420\ \Omega\right)\left[1 + \alpha'\left(T - T_0\right)\right]}{\left[1 + 2\alpha'\left(T - T_0\right)\right]} = \frac{\left(1.420\ \Omega\right)\left[1 + \left(17 \times 10^{-6}\ °C^{-1}\right)\left(80.0°C\right)\right]}{\left[1 + 2\left(17 \times 10^{-6}\ °C^{-1}\right)\left(80.0°C\right)\right]}$$

or $R = 1.418\ \Omega$
◊

Note: Some rules for handing significant figures have been deliberately violated in this solution in order to illustrate the very small difference in the results obtained with these two expressions.

Chapter 18
Direct-Current Circuits

NOTES FROM SELECTED CHAPTER SECTIONS

18.1   Sources of EMF

The source of emf (for example a battery or generator) maintains the constant current in a closed circuit. The emf, $\mathcal{E}$, of the source is the work done per unit charge in increasing the electric potential energy of the circulating charges. *One joule of work is required to move one coulomb of charge between two points which differ in potential by one volt.*

The terminal voltage (ΔV between the positive and negative terminals of a battery or other source) is equal to the emf of the battery when the current in the circuit is zero. In this case the terminal voltage is called the *open circuit voltage*. When the battery is delivering current to a circuit, the terminal voltage (*closed circuit voltage*) is less than the emf. This is due to the internal resistance of the battery.

18.2   Resistors in Series

For a group of resistors connected in series:

*   The equivalent resistance of the series combination is the algebraic sum of the individual resistances. See Equation 18.4.

*   The current is the same for each resistor in the group.

*   The total potential difference across the group of resistors equals the sum of the potential differences across the individual resistors.

18.3   Resistors in Parallel

For a group of resistors connected in parallel:

*   The reciprocal of the equivalent resistance of the group is the algebraic sum of the reciprocals of the individual resistors. See Equation 18.6.

- The potential differences across the individual resistors have the same value.

- The total current associated with the parallel group is the sum of the currents in the individual resistors.

18.4   Kirchhoff's Rules and Complex DC Circuits

When all resistors in a circuit are connected in series and/or parallel combinations, Ohm's law and the properties stated above in Sections 18.2 and 18.3 are adequate to analyze the circuit to determine the current in each resistor and the potential difference between any two points in the circuit.

When circuits are formed so that they cannot be reduced to a single equivalent resistor, it is necessary to use Kirchhoff's Rules to analyze the circuit.

1. **Junction rule (conservation of charge):** The sum of the currents entering any junction must equal the sum of the currents leaving that junction. (A junction is any point in the circuit where the current can split.)

2. **Loop rule (conservation of energy):** The algebraic sum of the changes in potential around any closed circuit loop must be zero.

18.5   *RC* Circuits

When the switch in a dc circuit containing only a battery and resistors is moved from the "open" to the "closed" position, the current in the circuit will have a constant value.

Consider a battery, capacitor and resistor in series as shown at right. When the switch is moved from the "open" to the "closed" position (the capacitor will begin charging), the charge on the capacitor increases with time, and the current decreases exponentially. Also, during one time constant, the charging current decreases from its initial maximum value of $I_0 = \mathcal{E}/R$ to approximately 37 percent of I_0. Note that $Q = C\mathcal{E}$ is the charge that would be accumulated on the capacitor if time, t, is allowed to go to infinity.

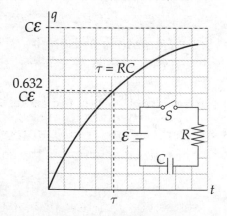

Charging capacitor

When the switch in a circuit containing an initially charged capacitor and a resistor is moved from the "open" to the "closed" position (the capacitor will begin discharging) the charge on the capacitor, current in the circuit, and the voltage across the resistor will all decrease exponentially with time.

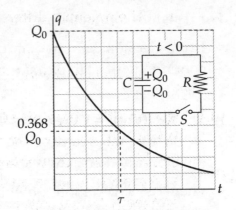

Discharging capacitor

In the process of charging or discharging a capacitor, the potential difference between the plates changes as charges are transferred from one plate of the capacitor to the other. *The transfer of charge produces a current in the circuit; the charges do not move across the gap between the plates of the capacitor.*

EQUATIONS AND CONCEPTS

The **terminal voltage of a battery** will be less than the emf when the battery is providing a current to an external circuit. This is due to **internal resistance** of the battery. *Often the internal resistance, r, of the source is very small compared to the external load resistance, R, of the circuit, and r can be neglected.*

$$\Delta V = \mathcal{E} - Ir \tag{18.1}$$

$$I = \frac{\mathcal{E}}{R + r}$$

For resistors connected in series:

- There is only one common circuit point per pair.

- The total or equivalent resistance of a series combination of resistors is equal to the sum of the resistances of the individual resistors.

$$R_{eq} = R_1 + R_2 + R_3 + \ldots \tag{18.4}$$

- Each resistor has the same current.

$$I_1 = I_2 = I_3 = \ldots$$

- The total potential difference equals the sum of the individual potential differences.

$$\Delta V = \Delta V_1 + \Delta V_2 + \Delta V_3 + \ldots$$

For resistors connected in parallel:

- Each resistor has two circuit points in common with each of the other resistors in the group.

- The reciprocal of the equivalent resistance is the algebraic sum of the reciprocals of the individual resistances.

$$\frac{1}{R_{eq}} = \frac{1}{R_1} + \frac{1}{R_2} + \frac{1}{R_3} \cdots \qquad (18.6)$$

- There is a common potential difference across the group of resistors.

$$\Delta V = \Delta V_1 = \Delta V_2 = \Delta V_3 = \cdots$$

- The total current equals the sum of the currents in the individual resistors.

$$I_{total} = I_1 + I_2 + I_3 + \cdots$$

In a **simple RC circuit** (capacitor, resistor, and battery in series) the charge on the capacitor increases as given by Eq.18.7.

$$q = Q\left(1 - e^{-t/RC}\right) \qquad (18.7)$$

The **time constant** of the circuit, τ represents the time required for the charge on the capacitor to reach 63.2% of maximum value. *In Equation 18.7, a large value of C will result in a larger final charge on the capacitor and a large value of R will limit the charging current. Large values of C or R imply a longer time required to achieve a given fraction of final charge.*

$$\tau = RC \qquad (18.8)$$

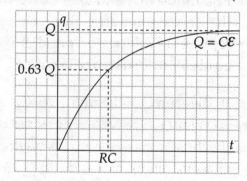

When a **capacitor is discharged** through a resistor, the charge and current will decrease exponentially with time.

$$q = Q_0 e^{-t/RC} \qquad (18.9)$$

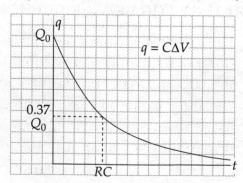

SUGGESTIONS, SKILLS, AND STRATEGIES

PROBLEM-SOLVING STRATEGY FOR RESISTORS

1. When two or more unequal resistors are connected in series, they carry the same current, but the potential differences across each of them are not the same. The resistances add directly to give the equivalent resistance of the series combination.

2. When two or more resistors (whether or not of equal value) are connected in parallel, the potential differences across them are the same. Since the current is inversely proportional to the resistance, the currents through them are not the same (except when the resistors are of equal value). *The equivalent resistance of a parallel combination of resistors is found through reciprocal addition, and the equivalent resistance is always less than the smallest individual resistance.*

3. A circuit consisting of several resistors, in series-parallel combination, can often be reduced to a simple circuit containing only one equivalent resistor. To do so, examine the initial circuit and replace any resistors in series or any in parallel using Equations 18.4 and 18.6 along with the properties outlined in Steps 1 and 2. Sketch the new circuit after these changes have been made. Examine the new circuit and replace any series or parallel combinations. **Continue this process until a single equivalent resistance is found.**

4. If the current through, or the potential difference across, a resistor in a circuit consisting of resistors in series-parallel combination is to be identified, start with the final circuit found in Step 3 and gradually work your way back through the circuits, using $\Delta V = IR$ and the properties of Steps 1 and 2. Record calculated values of current and potential difference on the circuit sketches as you proceed.

STRATEGY FOR USING KIRCHHOFF'S RULES

1. First, draw the circuit diagram and assign labels and symbols to all the known and unknown quantities. **You must assign directions to the currents in each branch of the circuit.** Do not be alarmed if you guess the direction of a current incorrectly; the resulting value will be negative, but its magnitude will be correct. Although the assignment of current directions is arbitrary, you must stick with your choice throughout as you apply Kirchhoff's rules.

2. Apply the junction rule to any junction in the circuit. The junction rule may be applied as many times as a new current (one not used in a previous application) appears in the resulting equation. **In general, the number of times the junction rule can be used is one fewer than the number of junction points in the circuit.**

3. Now apply Kirchhoff's loop rule to as many loops in the circuit as are needed to solve for the unknowns. Remember you must have as many equations as there are unknowns (I's, R's, and $\mathcal{E}$'s). **For each loop, you must first choose a direction to sum the voltage changes, clockwise or counterclockwise.** Then, you must correctly identify the increase or decrease in potential as you cross each resistor or source of emf in the closed loop.

4. Finally, you must solve the equations simultaneously for the unknown quantities. Be careful in your algebraic steps, and check your numerical answers for consistency.

Rules which should be used in determining the increase or decrease in potential difference as resistors or sources of emf are crossed in a circuit loop are stated below. In the accompanying figures, assume that the direction of travel on the section of circuit is from point "a" to point "b" (left to right).

1. The potential decreases (changes by $-IR$) when a resistor is traversed in the direction of the current.

    $$\Delta V = V_b - V_a = -IR$$

2. The potential increases (changes by $+IR$) when a resistor is traversed in the direction opposite the direction of the current.

    $$\Delta V = V_b - V_a = +IR$$

3. The potential increases by $+\mathcal{E}$ when a source of emf is traversed in the direction of the emf (from $-$ to $+$).

    $$\Delta V = V_b - V_a = \mathcal{E}$$

4. The potential decreases by $-\mathcal{E}$ when a source of emf is traversed opposite to the direction of the emf (from $+$ to $-$).

    $$\Delta V = V_b - V_a = -\mathcal{E}$$

An illustration of the use of Kirchhoff's rules for a three-loop circuit. In this illustration, the actual circuit elements, R's and $\mathcal{E}$'s, are not shown but assumed to be known. There are six possible different values of I in the circuit; therefore you will need six independent equations to solve for the six values of I. There are four junction points in the circuit (at points a, d, f, and h). The first rule applied at **any three of these points** will yield three equations. The circuit can be thought of as a group of three "blocks" as shown in the figure. Kirchhoff's second law, when applied to each of these loops (*abcda*, *ahfga*, and *defhd*), will yield three additional equations. You can then solve the total of six equations simultaneously for the six values of I_1, I_2, I_3, I_4, I_5, and I_6 (or some combination of currents, resistances and emf's totaling six quantities).

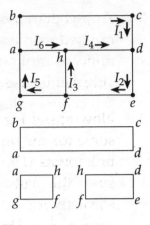

The sum of the changes in potential difference around any other closed loop in the circuit will be zero (for example, *abcdefga* or *ahfedcba*); however equations found by applying Kirchhoff's second rule to these additional loops will not be independent of the six equations found previously.

REVIEW CHECKLIST

▷ Calculate the current in a single-loop circuit and the potential difference between any two points in the circuit and calculate the equivalent resistance of a group of resistors in parallel, series, or series-parallel combination.

▷ Use Ohm's law to calculate the current in a circuit and the potential difference between any two points in a circuit which can be reduced to an equivalent single-loop circuit.

▷ Use Joule's law to calculate the power dissipated by any resistor or group of resistors in a circuit.

▷ Apply Kirchhoff's rules to solve multiloop circuits; that is, find the currents at any point and the potential difference between any two points.

▷ Describe in qualitative terms the manner in which charge accumulates on a capacitor or current flow changes through a resistor with time in a series circuit with battery, capacitor, resistor, and switch. Use Equations 18.7 and 18.9 to calculate q at any time t and also to find the time t for which the ratio q/Q_0 will have a specified value.

SOLUTIONS TO SELECTED END-OF-CHAPTER PROBLEMS

1. A battery having an emf of 9.00 V delivers 117 mA when connected to a 72.0-Ω load. Determine the internal resistance of the battery.

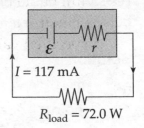

$I = 117$ mA

$R_{\text{load}} = 72.0$ W

Solution

The internal resistance, r, of the battery is in series with the load resistance, R_{load}. Thus, the total resistance of the circuit is

$$R_{eq} = r + R_{\text{load}} = r + 72.0 \ \Omega$$

Since a battery with emf $\mathcal{E} = 9.00$ V delivers 117 mA to the circuit, Ohm's law gives

$$R_{eq} = r + 72.0 \ \Omega = \frac{\mathcal{E}}{I} = \frac{9.00 \ \text{V}}{117 \times 10^{-3} \ \text{A}}$$

The internal resistance is therefore

$$r = \frac{9.00 \ \text{V}}{117 \times 10^{-3} \ \text{A}} - 72.0 \ \Omega = 4.92 \ \Omega \qquad \Diamond$$

5. (a) Find the equivalent resistance between points a and b in Figure P18.5. (b) Calculate the current in each resistor if a potential difference of 34.0 V is applied between points a and b.

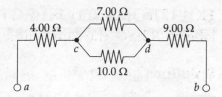

Figure P18.5

Solution

(a) The 7.00 Ω and the 10.0 Ω resistors between points c and d in the upper sketch are connected in parallel with each other. Thus, the total resistance between these points is

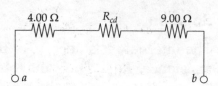

$$R_{cd} = \left(\frac{1}{7.00\ \Omega} + \frac{1}{10.0\ \Omega} \right)^{-1} = 4.12\ \Omega$$

This resistance is in series with the 4.00 Ω and the 9.00 Ω resistors as shown in the lower sketch. The total resistance between points a and b is then

$$R_{ab} = R_4 + R_{cd} + R_9 = 4.00\ \Omega + 4.12\ \Omega + 9.00\ \Omega = 17.1\ \Omega \qquad \lozenge$$

(b) If the potential difference between points a and b is $\Delta V_{ab} = 34.0$ V, the current from one of these points to the other is

$$I_{ab} = \frac{\Delta V_{ab}}{R_{ab}} = \frac{34.0\ \text{V}}{17.1\ \Omega} = 1.99\ \text{A}$$

This total current exists in both the 4.00 Ω and the 9.00 Ω resistors, so

$$I_4 = I_9 = 1.99\ \text{A} \qquad \lozenge$$

Also, $I_{cd} = I_{ab}$, so the potential difference between points c and d will be

$$\Delta V_{cd} = R_{cd}I_{ab} = (4.12\ \Omega)(1.99\ \text{A}) = 8.18\ \text{V}$$

The current in the 7.00 Ω resistor is then $I_7 = \dfrac{\Delta V_{cd}}{R_7} = \dfrac{8.18\ \text{V}}{7.00\ \Omega} = 1.17\ \text{A} \qquad \lozenge$

and that in the 10.0 Ω resistor is $I_{10} = \dfrac{\Delta V_{cd}}{R_{10}} = \dfrac{8.18\ \text{V}}{10.0\ \Omega} = 0.818\ \text{A} \qquad \lozenge$

Observe that $I_7 + I_{10} = I_{ab}$ as it should.

11. The resistance between terminals a and b in Figure P18.11 is 75 Ω. If the resistors labeled R have the same value, determine R.

Solution

The figures given below, show the steps in simplifying the original circuit by using the rule for combining resistors in series and parallel.

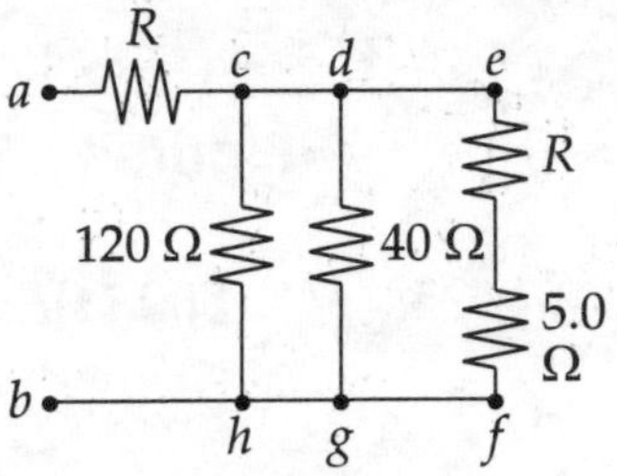

Figure 1 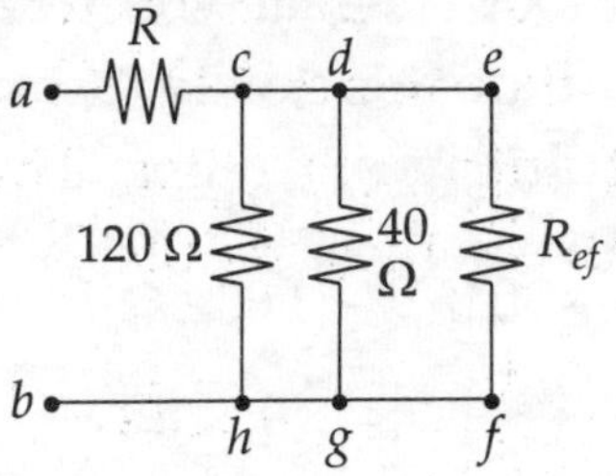   **Figure 2** 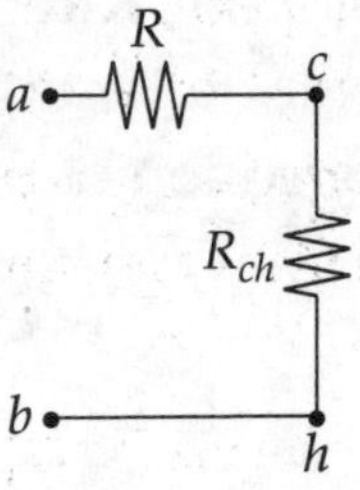   **Figure 3**

In Figure 1, observe that the two resistors between points e and f are in series. We replace this series combination by the single resistor $R_{ef} = R + 5.0\ \Omega$ as shown in Figure 2. Then, from Figure 2, we observe that the three resistances R_{ef}, 40 Ω, and 120 Ω are all connected in parallel. Recalling that $R_{ef} = R + 5.0\ \Omega$, the total resistance R_{ch} between points c and h is calculated using the rule for parallel resistors as

$$\frac{1}{R_{ch}} = \frac{1}{R+5.0\ \Omega} + \frac{1}{40\ \Omega} + \frac{1}{120\ \Omega} = \frac{4R+140\ \Omega}{(120\ \Omega)(R+5.0\ \Omega)} = \frac{R+35\ \Omega}{(30\ \Omega)(R+5.0\ \Omega)}$$

or $\quad R_{ch} = \dfrac{(30\ \Omega)(R+5.0\ \Omega)}{R+35\ \Omega}$

Then, looking at Figure 3, observe that the total resistance between points a and b is $R_{ab} = R + R_{ch}$. Thus, if $R_{ab} = 75\ \Omega$, we must have

$$75\ \Omega = R + \frac{(30\ \Omega)(R+5.0\ \Omega)}{R+35\ \Omega} = \frac{R^2 + (65\ \Omega)R + 150\ \Omega^2}{R+35\ \Omega}$$

which reduces to $R^2 - (10\ \Omega)R - 2\ 475\ \Omega^2 = 0 \quad$ or $\quad (R-55\ \Omega)(R+45\ \Omega) = 0$

Only the positive solution is physically acceptable, so $R = 55\ \Omega$ ◊

19. Figure P18.19 shows a circuit diagram. Determine (a) the current, (b) the potential of wire A relative to ground, and (c) the voltage drop across the 1 500-Ω resistor.

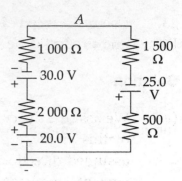

Figure P18.19

Solution

(a) Assume the current is clockwise around this circuit. Applying Kirchhoff's loop rule, as you start at the lower left corner and go clockwise around the single loop of this circuit, gives

$$+20.0 \text{ V} - (2\,000 \ \Omega)I - 30.0 \text{ V} - (1\,000 \ \Omega)I - (1\,500 \ \Omega)I + 25.0 \text{ V} - (500 \ \Omega)I = 0$$

or

$$I = \frac{(+20.0 - 30.0 + 25.0) \text{ V}}{(2\,000 + 1\,000 + 1\,500 + 500) \ \Omega} = \frac{+15.0 \text{ V}}{5\,000 \ \Omega} = 3.00 \times 10^{-3} \text{ A} = 3.00 \text{ mA} \qquad \Diamond$$

Since the computed current is positive, it is clockwise as assumed.

(b) Start at the lower left corner (i.e., the grounded point) and move up the left side of the circuit to conductor A at the top, recording changes in potential as you go. Taking the potential of the grounded point to be zero, this yields

$$V_A - 0 = +20.0 \text{ V} - (2\,000 \ \Omega)(3.00 \times 10^{-3} \text{ A}) - 30.0 \text{ V} - (1\,000 \ \Omega)(3.00 \times 10^{-3} \text{ A})$$

or the potential of conductor A is $\qquad V_A = -19.0 \text{ V} \qquad \Diamond$

(c) The change in potential across the 1 500-Ω resistor has a magnitude of

$$(\Delta V)_{1\,500} = (1\,500 \ \Omega)(3.00 \times 10^{-3} \text{ A}) = 4.50 \text{ V} \qquad \Diamond$$

Since the current is clockwise around this loop, it is directed downward through the 1 500-Ω resistor. Thus, the upper end of the 1 500-Ω resistor is at a higher electrical potential than the lower end.

23. Using Kirchhoff's rules, (a) find the current in each resistor shown in Figure P18.23 and (b) find the potential difference between points *c* and *f*.

Solution

(a) We assume the currents in the three branches of this circuit are directed as shown. If our assumed directions are correct, the calculated currents will have positive signs.

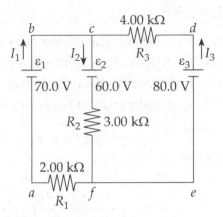

Figure P18.23

Apply Kirchhoff's junction rule at point *c* to obtain

$$I_2 = I_1 + I_3 \qquad [1]$$

Apply Kirchhoff's loop rule, going clockwise, around loop *abcdefa*:

$$70.0 \text{ V} + (4\,000\ \Omega)I_3 - 80.0 \text{ V} - (2\,000\ \Omega)I_1 = 0$$

or $\quad I_1 = (2.00)I_3 - 5.00\times10^{-3} \text{ A} \qquad [2]$

Apply Kirchhoff's loop rule, going counterclockwise, around loop *edcfe*:

$$80.0 \text{ V} - (4\,000\ \Omega)I_3 - 60.0 \text{ V} - (3\,000\ \Omega)I_2 = 0$$

or $\quad I_3 = 5.00\times10^{-3} \text{ A} - (0.750)I_2 \qquad [3]$

Solving Equations [1], [2], and [3] simultaneously yields

$$I_1 = 0.385 \text{ mA}, \ I_2 = 3.08 \text{ mA}, \text{ and } I_3 = 2.69 \text{ mA} \qquad \Diamond$$

All calculated currents are positive, so they are all directed as assumed and shown in the circuit diagram given above. $\qquad \Diamond$

(b) To determine the potential difference between points *c* and *f*, we start at point *f* and go up the center branch of the circuit to point *c*, recording all changes in potential that occur. This gives:

$$\Delta V_{cf} = V_c - V_f = (3.00 \text{ k}\Omega)I_2 + 60.0 \text{ V} = (3\,000\ \Omega)(3.08\times10^{-3} \text{ A}) + 60.0 \text{ V}$$

or $\quad \Delta V_{cf} = +69.2 \text{ V}$ (with point *c* at a higher potential than point *f*) $\qquad \Diamond$

29. Find the potential difference across each resistor in Figure P18.29.

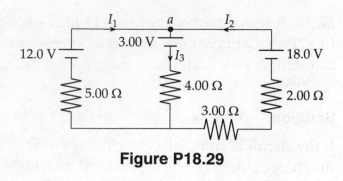

Figure P18.29

Solution

Assume that currents I_1, I_2, and I_3 are in the directions indicated by the arrows in the figure. Applying Kirchhoff's junction rule at junction a gives

$$I_3 = I_1 + I_2 \qquad \qquad \text{[1]}$$

Using Kirchhoff's loop rule as you start at point a and go clockwise around the leftmost loop yields

$$-3.00\ \text{V} - (4.00\ \Omega)I_3 - (5.00\ \Omega)I_1 + 12.0\ \text{V} = 0$$

or $\quad 5I_1 + 4I_3 = 9.00\ \text{A} \qquad \qquad \text{[2]}$

Starting at point a and going counterclockwise around the rightmost loop, Kirchhoff's loop rule gives

$$-3.00\ \text{V} - (4.00\ \Omega)I_3 - (3.00\ \Omega + 2.00\ \Omega)I_2 + 18.0\ \text{V} = 0$$

or $\quad 5I_2 + 4I_3 = 15.0\ \text{A} \qquad \qquad \text{[3]}$

Solving Equations [1], [2], and [3] simultaneously gives

$$I_1 = 0.323\ \text{A}, \quad I_2 = 1.523\ \text{A}, \text{ and } \quad I_3 = 1.846\ \text{A}$$

Therefore, the potential differences across the resistors are

$$\Delta V_2 = I_2(2.00\ \Omega) = 3.05\ \text{V} \qquad\qquad \Delta V_3 = I_2(3.00\ \Omega) = 4.57\ \text{V} \qquad \Diamond$$

$$\Delta V_4 = I_3(4.00\ \Omega) = 7.38\ \text{V} \qquad\qquad \Delta V_5 = I_1(5.00\ \Omega) = 1.62\ \text{V} \qquad \Diamond$$

34. A series combination of a 12-kΩ resistor and an unknown capacitor is connected to a 12-V battery. One second after the circuit is completed, the voltage across the capacitor is 10 V. Determine the capacitance of the capacitor.

Solution

If the circuit is completed by closing switch S at time $t = 0$, the charge stored on the capacitor at time t later is given by

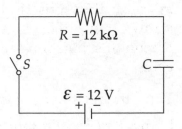

$$q = Q\left(1 - e^{-t/\tau}\right) \quad \text{where} \quad Q = C\mathcal{E} \quad \text{and the time}$$

constant is $\tau = RC$.

Thus, the potential difference across the capacitor at this time is

$$\Delta V = \frac{q}{C} = \frac{C\mathcal{E}\left(1 - e^{-t/\tau}\right)}{C} = \mathcal{E}\left(1 - e^{-t/\tau}\right)$$

If it is observed that $\Delta V = 10$ V at $t = 1.0$ s, then

$$10 \text{ V} = \left(12 \text{ V}\right)\left(1 - e^{-(1.0 \text{ s})/\tau}\right) \quad \text{or} \quad e^{-(1.0 \text{ s})/\tau} = 1 - \frac{10 \text{ V}}{12 \text{ V}} = \frac{1}{6}$$

Thus, $e^{+(1.0 \text{ s})/\tau} = 6$ giving $\dfrac{1.0 \text{ s}}{\tau} = \ln 6$ or $\tau = \dfrac{1.0 \text{ s}}{\ln 6}$

Since the time constant is $\tau = RC$, we find the unknown capacitance to be

$$C = \frac{\tau}{R} = \frac{1.0 \text{ s}}{\left(12 \times 10^3 \ \Omega\right)\ln 6} = 4.7 \times 10^{-5} \text{ F} = 47 \ \mu\text{F} \qquad \Diamond$$

39.   A heating element in a stove is designed to dissipate 3 000 W when connected to 240 V. (a) Assuming that the resistance is constant, calculate the current in the heating element if it is connected to 120 V. (b) Calculate the power it dissipates at that voltage.

Solution

The power dissipated in the heating element is

$$\mathcal{P} = (\Delta V)I = (\Delta V)\left(\frac{\Delta V}{R}\right) = \frac{(\Delta V)^2}{R}$$

Thus, the resistance of the heating element in normal operating conditions is

$$R = \frac{(\Delta V)^2}{\mathcal{P}} = \frac{(240 \text{ V})^2}{3\,000 \text{ W}} = 19.2 \ \Omega$$

(a)   If the resistance of the element is assumed to remain constant, the current when it is connected to a 120-V source will be

$$I' = \frac{\Delta V'}{R} = \frac{120 \text{ V}}{19.2 \ \Omega} = 6.25 \text{ A} \qquad \qquad \Diamond$$

(b)   The power dissipated when connected to the 120-V source would be

$$\mathcal{P}' = (\Delta V')I' = (120 \text{ V})(6.25 \text{ A}) = 750 \text{ W} \qquad \qquad \Diamond$$

Note that we could have also computed this power as

$$\mathcal{P}' = (I')^2 R = (6.25 \text{ A})^2 (19.2 \ \Omega) = 750 \text{ W} \qquad \qquad \Diamond$$

45. Find the equivalent resistance between points a and b in Figure P18.45.

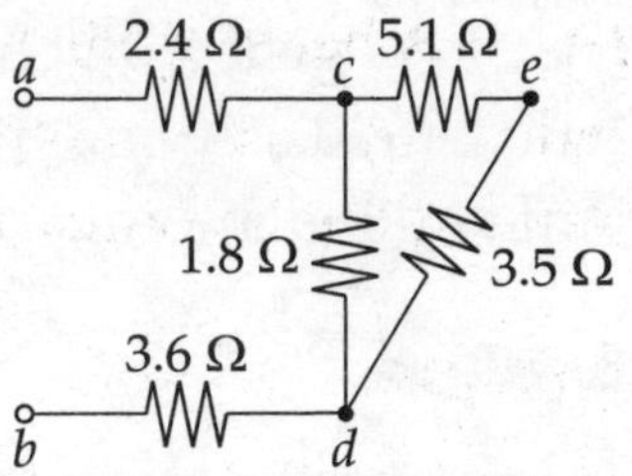

Figure P18.45

Solution

First, we observe that the path *ced* in the above figure contains a series combination of 2 resistors. The equivalent resistance of this combination is

$$R_{ced} = R_{ce} + R_{ed} = 5.1\ \Omega + 3.5\ \Omega = 8.6\ \Omega$$

Replacing this combination with its equivalent resistance, the circuit now looks like the figure at the right. Now, observe that there is a parallel combination of 2 resistors between points c and d. The equivalent resistance of this combination is

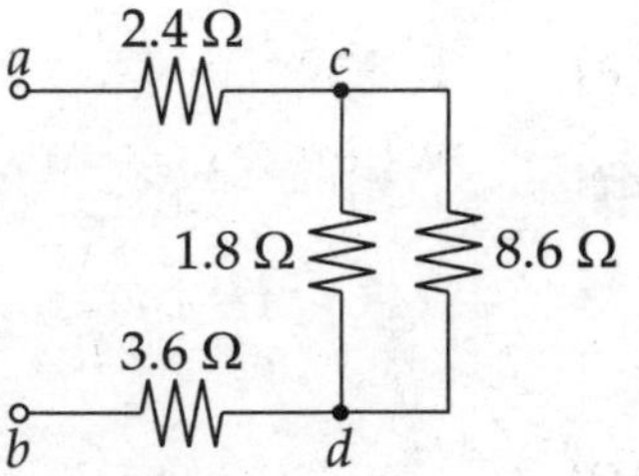

$$\frac{1}{R_{cd}} = \frac{1}{1.8\ \Omega} + \frac{1}{8.6\ \Omega} = \frac{8.6\ \Omega + 1.8\ \Omega}{(1.8\ \Omega)(8.6\ \Omega)} \quad \text{or} \quad R_{cd} = \frac{(1.8\ \Omega)(8.6\ \Omega)}{8.6\ \Omega + 1.8\ \Omega} = 1.5\ \Omega$$

Replacing this combination by its equivalent resistance, our equivalent circuit now shows 3 resistors in series between points a and b as shown at the right. Thus, the equivalent resistance between points a and b is

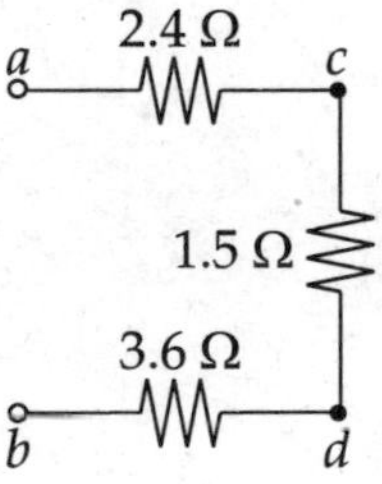

$$R_{ab} = R_{ac} + R_{cd} + R_{db}$$

or

$$R_{ab} = 2.4\ \Omega + 1.5\ \Omega + 3.6\ \Omega = 7.5\ \Omega$$

◊

53. A generator has a terminal voltage of 110 V when it delivers 10.0 A and 106 V when it delivers 30.0 A. Calculate the emf and the internal resistance of the generator.

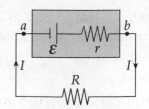

Solution

A generator, battery, or other power source may be considered to consist of a seat of emf, $\mathcal{E}$, and internal resistance r as shown within the shaded outline above. When this source supplies current I to a load, the voltage between the terminals of the source is

$$\Delta V_{ab} = \mathcal{E} - Ir$$

If the terminal voltage is $\Delta V_{ab} = 110$ V when the source delivers a current of $I = 10.0$ A, then

$$110 \text{ V} = \mathcal{E} - (10.0 \text{ A})r \qquad\qquad \textbf{[1]}$$

If it is also found that $\Delta V_{ab} = 106$ V when the current supplied to the load is $I = 30.0$ A, one has:

$$106 \text{ V} = \mathcal{E} - (30.0 \text{ A})r \qquad\qquad \textbf{[2]}$$

Subtracting Equation [2] from Equation [1] gives $\qquad 4 \text{ V} = (20.0 \text{ A})r$

so the internal resistance of the source is $\quad r = \dfrac{4 \text{ V}}{20.0 \text{ A}} = 0.2\ \Omega$ ◊

Then, Equation [1] gives the emf as $\qquad \mathcal{E} = 110 \text{ V} + (10.0 \text{ A})(0.2\ \Omega)$

$$\mathcal{E} = 112 \text{ V} \qquad\qquad\qquad ◊$$

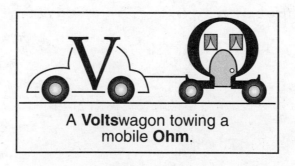

A **Volts**wagon towing a
mobile **Ohm**.

60. The circuit in Figure P18.60 contains two resistors, $R_1 = 2.0$ kΩ and $R_2 = 3.0$ kΩ, and two capacitors, $C_1 = 2.0$ μF and $C_2 = 3.0$ μF, connected to a battery with emf $\mathcal{E} = 120$ V. If there are no charges on the capacitors before switch S is closed, determine the charges q_1 and q_2 on capacitors C_1 and C_2, respectively, as functions of time after the switch is closed. [*Hint*: First reconstruct the circuit so that it becomes a simple RC circuit containing a single resistor and single capacitor in series, connected to the battery, and then determine the total charge q stored in the circuit.]

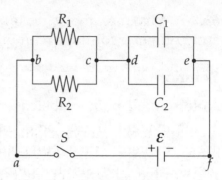

Figure P18.60

Solution

The equivalent resistance of the parallel combination of resistors between b and c is

$$\frac{1}{R_{bc}} = \frac{1}{2.0 \text{ k}\Omega} + \frac{1}{3.0 \text{ k}\Omega} = \frac{5.0}{6.0 \text{ k}\Omega} \quad \text{or} \quad R_{bc} = \frac{6.0 \text{ k}\Omega}{5.0} = 1.2 \text{ k}\Omega$$

Also, the total capacitance between points d and e is $\quad C_{de} = C_1 + C_2 = 5.0$ μF

so the time constant for this circuit is

$$\tau = R_{bc} C_{de} = \left(1.2 \times 10^3 \ \Omega\right)\left(5.0 \times 10^{-6} \ \text{F}\right) = 6.0 \times 10^{-3} \ \text{s} = 6.0 \text{ ms}$$

Thus, if the switch is closed at $t = 0$, the total charge stored between points d and e at time $t > 0$ is

$$q_{de} = Q\left(1 - e^{-t/\tau}\right) = C_{de}\mathcal{E}\left(1 - e^{-t/6.0 \text{ ms}}\right)$$

and the potential difference between d and e as a function of time is

$$\Delta V_{de} = \frac{q_{de}}{C_{de}} = \mathcal{E}\left(1 - e^{-t/6.0 \text{ ms}}\right) = (120 \text{ V})\left(1 - e^{-t/6.0 \text{ ms}}\right)$$

The charges on the capacitors C_1 and C_2 as functions of time are then

$$q_1 = C_1\left(\Delta V_{de}\right) = (2.0 \ \mu\text{F})(120 \text{ V})\left(1 - e^{-t/6.0 \text{ ms}}\right) = (240 \ \mu\text{C})\left(1 - e^{-t/6.0 \text{ ms}}\right) \qquad \Diamond$$

and $\quad q_2 = C_2\left(\Delta V_{de}\right) = (3.0 \ \mu\text{F})(120 \text{ V})\left(1 - e^{-t/6.0 \text{ ms}}\right) = (360 \ \mu\text{C})\left(1 - e^{-t/6.0 \text{ ms}}\right) \qquad \Diamond$

63. What are the expected readings of the ammeter and voltmeter for the circuit in Figure P18.63?

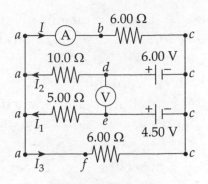

Figure P18.63

Solution

Note that all points labeled a in the circuit at the right are equivalent to each other and all the points labeled c are equivalent. The circuit could be redrawn with these merged into two single points a and c without altering the properties of the circuit.

To determine the voltmeter reading, we start at point e and follow the path ecd to point d, keeping track of all changes that occur in the potential. The result is

$$\Delta V_{de} = V_d - V_e = -4.50 \text{ V} + 6.00 \text{ V} = +1.50 \text{ V}$$

Thus, the voltmeter reads 1.50 V , with point d at the higher potential. ◊

Starting on the currents, we apply Kirchhoff's loop rule, going clockwise around loop $abcfa$ to obtain

$$-(6.00 \ \Omega)I + (6.00 \ \Omega)I_3 = 0 \qquad \text{or} \qquad I_3 = I \qquad \text{[1]}$$

Next, apply Kirchhoff's loop rule, going clockwise around loop $abcda$:

$$-(6.00 \ \Omega)I + 6.00 \text{ V} - (10.0 \ \Omega)I_2 = 0 \qquad \text{or} \qquad I_2 = 0.600 \text{ A} - 0.600 I \quad \text{[2]}$$

Applying Kirchhoff's loop rule to the loop $abcea$ gives

$$-(6.00 \ \Omega)I + 4.50 \text{ V} - (5.00 \ \Omega)I_1 = 0 \qquad \text{or} \qquad I_1 = 0.900 \text{ A} - 1.20 I \quad \text{[3]}$$

Finally, we apply Kirchhoff's junction rule at either point a or point c to obtain

$$I + I_3 = I_1 + I_2 \qquad \qquad \text{[4]}$$

Substituting Equations [1], [2], and [3] into Equation [4] gives the current through the ammeter and hence the ammeter reading. The result of this substitution is

$$I + I = 0.900 \text{ A} - 1.20 I + 0.600 \text{ A} - 0.600 I$$

or $\qquad 3.80 I = 1.50 \text{ A} \qquad$ and $\qquad I = \dfrac{1.50 \text{ A}}{3.80} = 0.395 \text{ A} = 395 \text{ mA}$ ◊

Chapter 19

Magnetism

NOTES FROM SELECTED CHAPTER SECTIONS

19.1 Magnets

19.2 Earth's Magnetic Field

Like magnetic poles (north-north or south-south) repel each other, and unlike poles (north-south) attract each other. Further, magnetic poles cannot be isolated and always occur in north-south pairs. **Soft magnetic materials**, such as iron, are easily magnetized but also tend to lose their magnetism easily; **hard magnetic materials** such as cobalt and nickel are difficult to magnetize but tend to retain their magnetism. The region of space surrounding any magnetic material can be characterized by a magnetic field represented by the symbol $\vec{B}$.

The magnetic field of the Earth can be pictured as having a pattern of magnetic field lines similar to the field lines surrounding a bar magnet. The Earth's geographic North Pole (in the Northern hemisphere) corresponds to a magnetic south pole. That is, the "north-seeking" end of a magnet compass needle will point toward the Northern hemisphere (but not directly toward the Earth's geographic North Pole).

At any given location on the surface of the Earth a compass needle will indicate two angles related to the Earth's magnetic field:

Dip angle: the angle between the horizontal, relative to the surface of the Earth, and the direction of the magnetic field at that point.

Magnetic declination: the angle between true North (geographic North pole) and magnetic North (indicated by a compass needle).

19.3   Magnetic Fields

The magnetic field is a vector quantity designated by the symbol $\vec{B}$ and has SI units of tesla (T). By convention the direction of the magnetic field vector is shown graphically as illustrated here. Magnetic field:

in the plane of the page is shown by arrows in the plane.

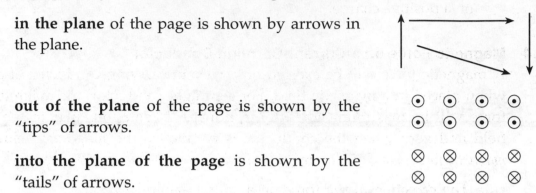

out of the plane of the page is shown by the "tips" of arrows.

into the plane of the page is shown by the "tails" of arrows.

Particles with charge q, moving with speed $\vec{v}$ in a magnetic field $\vec{B}$, experience a magnetic force $\vec{F}_B$. Properties of the magnetic force are:

- The magnetic force is proportional to the charge q and speed v of the particle.

- The magnitude of the magnetic force depends on the angle between the velocity vector of the particle and the magnetic field vector.

- When a charged particle moves in a direction parallel to the magnetic field vector, the magnetic force $\vec{F}_B$ on the charge is zero.

- The magnetic force acts in a direction perpendicular to both $\vec{v}$ and $\vec{B}$; that is, $\vec{F}_B$ is perpendicular to the plane formed by $\vec{v}$ and $\vec{B}$.

- The magnetic force on a positive charge is in the direction opposite to the force on a negative charge moving in the same direction.

- If the velocity vector makes an angle θ with the magnetic field, the magnitude of the magnetic force is proportional to $\sin\theta$.

The direction of the magnetic force on a **positive charge** can be determined by using **right-hand rule #1**:

1.   Point the fingers of your right hand in the direction of the velocity, $\vec{v}$.

2. Curl the fingers in the direction of the magnetic field, $\vec{B}$, moving through the smaller of the two possible angles.

3. Your thumb is now pointing in the direction of the magnetic force $\vec{F}$ exerted on a positive charge.

19.4 Magnetic Force on a Current-Carrying Conductor

A magnetic force will be exerted on a current-carrying conductor of any shape when placed in a magnetic field. For a straight conductor, the magnitude of the force will be maximum when the conductor is perpendicular to the magnetic field and zero when the conductor is parallel to the magnetic field. *The total magnetic force on a closed current loop in a uniform magnetic field is zero.*

The direction of the force on a straight conductor can be determined by using **right-hand rule #1**.

In this case you must place your fingers in the direction of the positive current, *I*, rather than in the direction of $\vec{v}$.

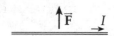

Can you use the right-hand rule to determine the direction of the magnetic field that exerts a force as shown on the conductor at the right?

19.5 Torque on a Current Loop and Electric Motors

As stated above, no net force is exerted on a closed current loop (coil) in a uniform magnetic field; however, there is a **net torque** exerted on a current loop in a magnetic field. The magnitude of the torque is proportional to the area of the loop, magnetic field strength, loop current, and the sine of the angle between the perpendicular to the plane of the loop and the direction of the magnetic field.

The torque on the loop will be maximum when the field is parallel to the plane of the loop, and the torque will be zero when the field is perpendicular to the plane of the loop. *The torque will cause the loop to rotate so that the normal to the plane of the loop will turn toward the direction parallel to the magnetic field.*

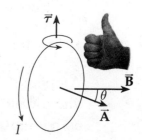

19.6   Motion of a Charged Particle in a Magnetic Field

A charged particle entering a uniform magnetic field with its velocity vector initially along a direction perpendicular to the field will move in a circular path in a plane perpendicular to the magnetic field. The magnetic force on the moving charge will be directed toward the center of the circular path and produce a centripetal acceleration. *The magnetic force changes the direction of the velocity vector but does not change its magnitude; the work done by the magnetic force on the particle is zero.*

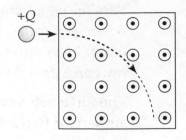

19.7   Magnetic Field of a Long, Straight Wire and Ampère's Law

A magnetic field will exist in the vicinity of a current-carrying wire. For a current in a long wire (the length of the wire is much greater than the distance from the wire) the magnetic field is inversely proportional to the distance from the conductor. The direction of the magnetic field due to a current in a long straight wire is given by **right-hand rule #2:**

> **If the wire is grasped in the right hand with the thumb in the direction of the positive current, the fingers will wrap (or curl) in the direction of $\vec{B}$.**

As illustrated in the figure, the magnetic field lines are circular in a plane perpendicular to the length of the conductor.

At any point in space surrounding the conductor, the magnetic field vector, $\vec{B}$, is directed along the tangent to the circle through that point.

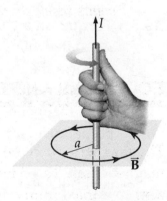

Ampère's circuital law states a method for calculating the magnetic field in the vicinity of a current-carrying conductor. *This technique is valid only for steady currents and can be easily used only in those cases where the current configuration has a high degree of symmetry.*

19.8 Magnetic Force Between Two Parallel Conductors

Parallel conductors carrying currents in the same direction attract each other, whereas parallel conductors carrying currents in opposite directions repel each other. *The forces on the two conductors are equal regardless of the relative value of the two currents.*

The force between two parallel wires each carrying a current is used to define the ampere. If two long, parallel wires 1 m apart in a vacuum carry the same current and the force per unit length on each wire is 2×10^{-7} N/m, then the current is defined to be 1 A.

19.9 Magnetic Fields of a Current Loops and Solenoids

The direction of the magnetic field at the center of a current loop is perpendicular to the plane of the loop and directed in the sense given by right-hand rule #2. For the situation illustrated in the figure, $\vec{B}$ is directed out of the page within the area of the loop and into the page outside the loop.

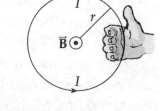

Within a solenoid, the magnetic field is parallel to the axis of the solenoid and pointing in a sense determined by applying right-hand rule #2 to one of the coils in the winding. *The magnetic field has a constant value at all points inside the solenoid (except near the ends).*

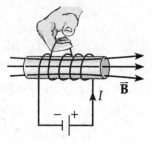

EQUATIONS AND CONCEPTS

The **magnetic force on a moving charge** in a magnetic field can have values ranging from zero to a maximum value, depending on the angle between the directions of the field, $\vec{B}$, and the velocity, $\vec{v}$, of the charge.

$$F = qvB\sin\theta \qquad (19.1)$$

The **maximum value** of the magnetic force occurs when the charge is moving perpendicular to the direction of the magnetic field.

$$F_{max} = qvB \qquad (19.4)$$

The **magnetic field intensity B** at some point in space is defined in terms of the magnetic force exerted on a moving positive electric charge at that point.

$$B \equiv \frac{F}{qv\sin\theta} \qquad (19.2)$$

The SI unit of magnetic field intensity is the tesla (T).

$$[B] = T = Wb/m^2 = N/A \cdot m \qquad (19.3)$$

Note in Equation 19.1 above, that when $\theta = 0$, the force, $F = 0$. *Therefore do not misinterpret Equation 19.2 to mean that B will be infinite when $\theta = 0$.*

The **magnetic force exerted on a current-carrying conductor** placed in a magnetic field depends on the angle between the direction of the current in the conductor and the direction of the field. The magnetic force will be maximum when the conductor is directed perpendicular to the magnetic field. *Equation 19.6 applies in the case of a straight conductor.*

$$F = BI\ell\sin\theta \qquad (19.6)$$

$$F_{max} = BI\ell \qquad (19.5)$$

The **magnitude of the torque on a current loop** in an external magnetic field depends on the area of the loop, the angle between the direction of the magnetic field, the direction of the normal (or perpendicular) to the plane of the loop, current, and magnitude of the magnetic field. For a multi-loop coil, the torque is proportional to the number of turns, N.

$$\tau = BIA\sin\theta \qquad (19.8)$$
(For a single loop)

θ = angle between the direction of $\vec{B}$ and the normal to the plane of the loop

Maximum magnitude of the torque will occur when the magnetic field is parallel to the plane of the loop.

$$\tau_{max} = BIA \qquad (19.7)$$

The **magnetic moment of a current loop**, $\vec{\mu}$, can be used to express the torque on the loop when placed in a magnetic field. The direction of $\vec{\mu}$ is perpendicular to the plane of the loop, and θ is the angle between the directions of $\vec{\mu}$ and $\vec{B}$.

$$\mu = IAN$$

$$\tau = \mu B\sin\theta \qquad (19.9b)$$

The **direction of rotation of the loop** decreases the angle between the normal to the loop and the magnetic field (the loop turns toward a position of minimum torque). The loop shown in the figure will rotate counterclockwise as seen from above.

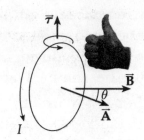

Motion of a charged particle entering a uniform magnetic field with the velocity vector initially perpendicular to the field has the following characteristics:

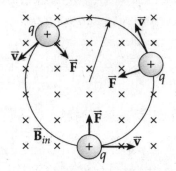

- The **path** of the particle will be circular and in a plane perpendicular to the direction of the field. The direction of rotation of the particle will be as determined by the right-hand rule.

- The radius of the circular path will be proportional to the linear momentum of the charged particle.

$$r = \frac{mv}{qB} \qquad (19.10)$$

- The period of revolution in the circular path described above is independent of the value of the radius.

$$T = \frac{2\pi r}{v} = \frac{2\pi m}{qB}$$

The magnetic fields due to the following current-carrying conductors have important applications. The value of the permeability constant is stated in Eq. 19.12.

$$\mu_0 = 4\pi \times 10^{-7} \text{ T·m/A} \qquad (19.12)$$

The magnitude of the magnetic field:

At a distance r from a **long straight conductor**.

$$B = \frac{\mu_0 I}{2\pi r} \qquad (19.11)$$

At the **center of a current loop** of N turns and radius R.

$$B = N\left(\frac{\mu_0 I}{2R}\right)$$

Inside a solenoid which has n turns of conductor per unit length (note that the value of B is independent of the radius of the solenoid).

$$B = \mu_0 n I \qquad (19.16)$$

N = total turns

n = turns per meter

Ampère's circuital law can be used to find the magnetic field around certain simple current-carrying conductors. The product $(B_{\parallel} \, \Delta\ell)$ must be summed around a closed path containing the current

$$\sum B_{\parallel} \, \Delta\ell = \mu_0 I \qquad (19.13)$$

$B_{\parallel}$ is the component of $\vec{B}$ tangent to a small displacement of length $\Delta\ell$

The **magnitude of the force, per unit length, between parallel conductors** is proportional to the product of the two currents and inversely proportional to d, the distance between the two conductors. Parallel currents in the same direction attract each other, and parallel currents in opposite directions repel each other. *The forces on the two conductors will be equal in magnitude regardless of the relative magnitude of the two currents.*

$$\frac{F_1}{\ell} = \frac{\mu_0 I_1 I_2}{2\pi d} \qquad (19.14)$$

REVIEW CHECKLIST

▷   Use right-hand rule #1 and appropriate equations to determine the direction and the magnitude of the magnetic force exerted on a moving electric charge or current-carrying conductor in a region where there is a magnetic field. Practice the right-hand rule for situations in which different quantities represent the unknown direction.

▷   Determine the magnitude of the torque exerted on a closed current loop in an external magnetic field and state the direction of rotation.

▷   Calculate the period and radius of the circular orbit of a charged particle moving in a uniform magnetic field.

▷   Calculate the magnitude and determine the direction (using right-hand rule #2) of the magnetic field for the following cases: (i) a distance r from a long, straight current-carrying conductor, (ii) at the center of a current loop of radius R, and (iii) at interior points of a solenoid with n turns per unit length.

▷   Use the superposition principle to calculate the net magnetic force in the vicinity of several current-carrying wires.

SOLUTIONS TO SELECTED END-OF-CHAPTER PROBLEMS

3. Find the direction of the magnetic field acting on the positively charged particle moving in the various situations shown in Figure P19.3 if the direction of the magnetic force acting on it is as indicated.

Figure P19.3

Solution

The magnetic force experienced by a moving charged particle is always perpendicular to both the velocity of the particle and the direction of the magnetic field. This means that the magnetic field, $\vec{B}$, must lie in the plane that is perpendicular to the force $\vec{F}$. However, without additional information, one cannot determine the exact orientation of $\vec{B}$ within that plane.

If we assume that the velocity, $\vec{v}$, of the particle is perpendicular to the magnetic field, right-hand rule #1 may be used to determine the direction of the field in each of the given situations.

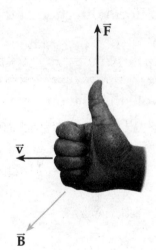

Right-hand rule #1 is illustrated in the drawing at the right. Normally, this is used to find the direction of the magnetic force exerted on a positively charged particle, given the directions of the field and particle's velocity. Alternatively, we may use it to find the direction of the magnetic field, $\vec{B}$, from the known directions of the force and the velocity.

For each of the situations given in this problem, hold your right hand with the fingers extended in the given direction of $\vec{v}$ and the thumb pointing in the direction of the force exerted on the positive charge. Then, after the fingers have moved 90° as you are closing your hand, they will point in the direction of the magnetic field . This technique yields the following results:

(a)   The field is directed into the page.   ◊

(b) $\vec{B}$ must be directed toward the right edge of the page.   ◊

(c)   The direction of the field is toward the bottom of the page.   ◊

7.   What velocity would a proton need to circle Earth 1 000 km above the magnetic equator, where Earth's magnetic field is directed horizontally north and has a magnitude of 4.00×10^{-8} T?

Solution

The gravitational force acting on the proton will be negligible in comparison to the magnetic force. Thus, the magnetic force must supply the needed centripetal acceleration of the proton. To do this, the magnetic force must be directed downward, toward Earth, and have a magnitude given by

$$F = qvB\sin\theta = \frac{mv^2}{r}$$

Here, q is the charge of the proton, v is its speed, θ is the angle the velocity makes with magnetic field $\vec{B}$, m is the mass of the proton, and the radius of the proton's orbit is

$$r = R_E + 1\,000 \text{ km} = 6.38 \times 10^6 \text{ m} + 1\,000 \times 10^3 \text{ m} = 7.38 \times 10^6 \text{ m}$$

Since the proton moves parallel to the equator, and hence perpendicular to the south-to-north direction of the magnetic field, $\theta = 90.0°$ and the required speed is

$$v = \frac{qrB}{m} = \frac{\left(1.60 \times 10^{-19} \text{ C}\right)\left(7.38 \times 10^6 \text{ m}\right)\left(4.00 \times 10^{-8} \text{ T}\right)}{1.67 \times 10^{-27} \text{ kg}} = 2.83 \times 10^7 \text{ m/s} \qquad \Diamond$$

Note that the proton must move east to west around its orbit if the magnetic force is to be directed downward. $\qquad \Diamond$

11. A current $I = 15$ A is directed along the positive x-axis and perpendicular to a magnetic field. A magnetic force per unit length of 0.12 N/m acts on the conductor in the negative y-direction. Calculate the magnitude and direction of the magnetic field in the region through which the current passes.

Solution

The current, flowing in the positive x direction is known to be perpendicular to the magnetic field. Also, the force that a magnetic field exerts on a current-carrying conductor is always perpendicular to the magnetic field. Thus, the field must be perpendicular to both the x-axis and the y-axis. This means that the field is parallel to the z-axis.

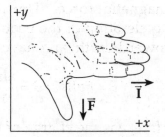

The direction that $\vec{\mathbf{B}}$ points along that axis can be determined by a modified version of right-hand rule #1. Holding your right hand with the fingers extended in the direction of the current (positive x direction) and the thumb pointing in the direction of the force (negative y direction), your fingers move toward the direction of $\vec{\mathbf{B}}$ as you close your hand. You should find this to be the positive z direction (that is, out of the page toward you). ◊

The magnitude of the magnetic field is found from $F = BI\ell\sin\theta$, with $\theta = 90°$ since the current is perpendicular to the field. This gives

$$B = \frac{F/\ell}{I\sin\theta} = \frac{0.12 \text{ N/m}}{(15 \text{ A})\sin 90°} = 8.0 \times 10^{-3} \text{ T} = 8.0 \text{ mT} \qquad ◊$$

17.   A wire with a mass of 1.00 g/cm is placed on a horizontal surface with a coefficient of friction of 0.200. The wire carries a current of 1.50 A eastward and moves horizontally to the north. What are the magnitude and the direction of the *smallest* vertical magnetic field that enables the wire to move in this fashion?

Solution

The smallest possible magnetic field which can move the wire in the desired manner will exert a horizontal force directed due northward (the direction we want the wire to move). Thus, $\vec{\mathbf{F}}$, $\vec{\mathbf{B}}$, and $\vec{\mathbf{I}}$ are all perpendicular to each other.

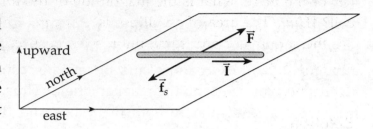

To determine whether $\vec{\mathbf{B}}$ must be vertically upward or downward, use the modified version of right-hand rule #1. Hold your right hand with the extended fingers pointing in the direction of the current in the wire (eastward) and the thumb pointing in the desired direction of the magnetic force (northward). As you close your hand, the fingers move toward the required direction of $\vec{\mathbf{B}}$. You should find that this is downward toward the ground.   ◊

To determine the required magnitude of the magnetic field, realize that if the magnetic force acting on the wire is horizontal, the normal force exerted on the wire by the horizontal surface is simply the weight, *mg*, of the wire. The maximum friction force retarding the motion of the wire is then

$$f_{s,\,max} = \mu_s n = \mu_s mg$$

When the magnetic field is just able to move the wire, the magnitude of the magnetic force equals the magnitude of the maximum friction force, or

$$F = f_{s,\,max} \quad \Rightarrow \quad BI\ell\sin 90.0° = \mu_s mg$$

With $m/\ell = 1.00$ g/cm $= 100$ g/m $= 0.100$ kg/m, this gives

$$B = \frac{\mu_s mg}{I\ell} = \frac{\mu_s (m/\ell)g}{I} = \frac{(0.200)(0.100 \text{ kg/m})(9.80 \text{ m/s}^2)}{1.50 \text{ A}} = 0.131 \text{ T} \qquad ◊$$

23. An eight-turn coil encloses an elliptical area having a major axis of 40.0 cm and a minor axis of 30.0 cm (Fig. P19.23). The coil lies in the plane of the page and has a 6.00-A current flowing clockwise around it. If the coil is in a uniform magnetic field of 2.00×10^{-4} T directed toward the left of the page, what is the magnitude of the torque on the coil? [*Hint*: The area of an ellipse is $A = \pi ab$, where a and b are the semimajor and semiminor axes, respectively, of the ellipse.]

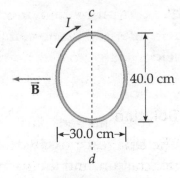

**Figure P19.23
(modified)**

Solution

The torque that a uniform magnetic field exerts on a flat (or planar coil) is given by

$$\tau = NBIA \sin \theta$$

where N is the number of turns of wire on the coil, B is the magnitude of the magnetic field, I is the current flowing in the coil, A is the area enclosed by the coil, and θ is the angle between the line perpendicular to the plane of the coil and the direction of the magnetic field. While the expression given above was derived for a rectangular coil in the textbook, it is valid for any planar coil. The area enclosed by the coil may be square, rectangular, circular, ellipsoidal, or any two-dimensional geometric shape.

In the situation shown in Figure P19.23, the magnetic field is parallel to the plane of the coil. Thus, the angle between the magnetic field and the line perpendicular to the plane of the coil is $\theta = 90.0°$. With semi-major and semi-minor axes of $a = 0.400$ m/2 and $b = 0.300$ m/2, the enclosed area of the coil is

$$A = \pi ab = \pi (0.200 \text{ m})(0.150 \text{ m}) = 0.030\pi \text{ m}^2$$

and the magnitude of the torque exerted on the coil is

$$\tau = 8(2.00 \times 10^{-4} \text{ T})(6.00 \text{ A})(0.030\pi \text{ m}^2)\sin 90.0° = 9.05 \times 10^{-4} \text{ N} \cdot \text{m} \qquad \lozenge$$

Note that the magnetic force exerted on the left side of the coil is directed out of the page while the force exerted on the right side is into the page. Thus, an observer looking from c toward d along the dashed line shown in the figure would see the coil rotate counterclockwise.

27. A proton moving freely in a circular path perpendicular to a constant magnetic field takes 1.00 μs to complete one revolution. Determine the magnitude of the magnetic field.

Solution

When a charged particle moves perpendicular to a constant magnetic field, the magnetic force exerted on the particle by the field supplies the needed centripetal acceleration, and the particle follows a circular path. That is,

$$F = qvB\sin 90° = m\frac{v^2}{r}$$

Thus, the speed of the particle is given by

$$v = \frac{qBr}{m}$$

and the time required to complete one revolution around the circular path is

$$T = \frac{circumference}{v} = \frac{2\pi r}{qBr/m} = \frac{2\pi m}{qB}$$

For a proton, $q = +e$ and $m = 1.67 \times 10^{-27}$ kg. If it is observed that $T = 1.00$ μs, the magnitude of the field must be

$$B = \frac{2\pi m}{eT} = \frac{2\pi \left(1.67 \times 10^{-27} \text{ kg}\right)}{\left(1.60 \times 10^{-19} \text{ C}\right)\left(1.00 \times 10^{-6} \text{ s}\right)} = 6.56 \times 10^{-2} \text{ T} \qquad \Diamond$$

32. A mass spectrometer is used to examine the isotopes of uranium. Ions in the beam emerge from the velocity selector at a speed of 3.00×10^5 m/s and enter a uniform magnetic field of 0.600 T directed perpendicularly to the velocity of the ions. What is the distance between the impact points formed on the photographic plate by singly charged ions of ^{235}U and ^{238}U?

Solution

Charged particles moving perpendicular to a uniform magnetic field experience a force that supplies the centripetal acceleration and deflects them into circular paths as shown at the right.

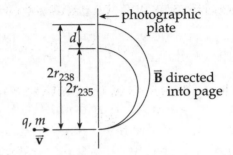

That is, $$F = qvB \sin 90° = m\frac{v^2}{r}$$

and the radius of the circular path is $$r = \frac{mv}{qB}$$

If the beam of particles entering the mass spectrometer with speed v contains ^{235}U and ^{238}U ions, both having net charge q but differing in mass, the distance between the impact points on the photographic plate is

$$d = 2r_{238} - 2r_{235} = 2(\Delta r) = \frac{2(\Delta m)v}{qB} = \frac{2\left[(238-235)\text{ u}\right]v}{qB} = \frac{(6\text{ u})v}{qB}$$

where the unified mass unit is $1\text{ u} = 1.66 \times 10^{-27}$ kg. For singly charged ions, $q = +1e$ and we find that

$$d = \frac{6\left(1.66 \times 10^{-27}\text{ kg}\right)\left(3.00 \times 10^5\text{ m/s}\right)}{\left(1.60 \times 10^{-19}\text{ C}\right)\left(0.600\text{ T}\right)} = 3.11 \times 10^{-2}\text{ m} = 3.11\text{ cm} \qquad \Diamond$$

41. A wire carries a 7.00-A current along the x-axis, and another wire carries a 6.00-A current along the y-axis, as shown in Figure P19.41. What is the magnetic field at point P, located at $x = 4.00$ m, $y = 3.00$ m?

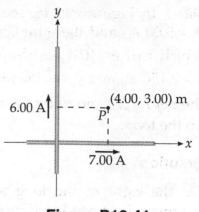

Figure P19.41

Solution

Let us call the wire along the x-axis wire 1 and that along the y-axis wire 2. Also, we shall consider a magnetic field at point P to be positive when directed out of the page toward the reader and negative if directed into the page.

The magnetic field at point P is the vector sum of the field generated by the current in wire 1 and the field generated by the current in wire 2. The magnitude of the field due to a current I in a long straight wire is given by

$$B = \frac{\mu_0 I}{2\pi r}$$

where r is the perpendicular distance from the observation point to the wire.

The direction of the field due to a long, straight, conductor may be determined by imagining your right hand gripping the conductor with the thumb pointing along the conductor in the direction of the current. When this is done, the fingers circle around the conductor in the direction of the magnetic field lines produced by the current. Applying this technique to the wire along the x-axis shows that the contribution by wire 1 to the field at point P is directed out of the page and hence, is positive. Using the same method for the wire 2 indicates that its contribution to the field at P is directed into the page (negative direction).

Therefore, the resultant field at P is given by $\vec{B} = \vec{B}_1 + \vec{B}_2$:

$$\vec{B} = +\frac{\mu_0 I_1}{2\pi r_1} - \frac{\mu_0 I_2}{2\pi r_2} = \frac{4\pi \times 10^{-7} \text{ T·m/A}}{2\pi}\left(\frac{7.00 \text{ A}}{3.00 \text{ m}} - \frac{6.00 \text{ A}}{4.00 \text{ m}}\right) = +1.67 \times 10^{-7} \text{ T}$$

or $\vec{B} = 0.167 \ \mu\text{T}$ directed out of the page   ◊

46. In Figure P19.46, the current in the long, straight wire is $I_1 = 5.00$ A, and the wire lies in the plane of the rectangular loop, which carries 10.0 A. The dimensions shown are $c = 0.100$ m, $a = 0.150$ m, and $\ell = 0.450$ m. Find the magnitude and direction of the net force exerted by the magnetic field due to the straight wire on the loop.

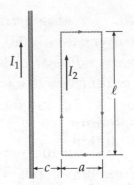

Figure P19.46

Solution

On the right of the long straight wire in Figure P19.46, the magnetic field produced by the current I_1 at distance r from the wire is directed into the page and has magnitude

$$B_1 = \frac{\mu_0 I_1}{2\pi r}$$

The force on a straight conductor of length ℓ and carrying current I_2 in this field is given by $F = B_1 I_2 \ell \sin\theta$, where θ is the angle between the direction of the current I_2 and the direction of the field B_1. For each part of the rectangular loop, $\theta = 90°$ and $\sin\theta = 1$. The direction of the force is given by right-hand rule #1.

Each segment of the upper end of the rectangular loop in Figure P19.46 experiences a force directed toward to the top of the page. For each such segment, there is a corresponding segment in the lower end of the loop that experiences an equal magnitude force directed toward the bottom of the page. Hence, the total forces on the ends of the loop cancel each other.

Then, the net force exerted on the loop by the magnetic field $\vec{B}_1$ is the vector sum of the forces on the left and right sides of the loop. Taking toward the right as positive and toward the long, straight wire as negative, the net force is

$$\vec{F}_{net} = \vec{F}_{\text{left side}} + \vec{F}_{\text{right side}} = -\left(B_1\big|_{r=c}\right)I_2\ell + \left(B_1\big|_{r=c+a}\right)I_2\ell = \frac{\mu_0 I_1}{2\pi}\left(-\frac{1}{c} + \frac{1}{c+a}\right)I_2\ell$$

or $\quad \vec{F}_{net} = \dfrac{\left(4\pi\times10^{-7}\ \text{T·m/A}\right)\left(5.00\ \text{A}\right)}{2\pi}\left(-\dfrac{1}{0.100\ \text{m}} + \dfrac{1}{0.250\ \text{m}}\right)\left(10.0\ \text{A}\right)\left(0.450\ \text{m}\right)$

$\vec{F}_{net} = -2.70\times10^{-5}$ N $\quad$ or $\quad \vec{F}_{net} = 2.70\times10^{-5}$ N $\quad$ toward the left $\qquad\qquad \Diamond$

49.   A single-turn square loop of wire 2.00 cm on a side carries a counterclockwise current of 0.200 A. The loop is inside a solenoid, with the plane of the loop perpendicular to the magnetic field of the solenoid. The solenoid has 30 turns per centimeter and carries a counterclockwise current of 15.0 A. Find the force on each side of the loop and the torque acting on the loop.

Solution

The magnetic field inside the solenoid is uniform and directed parallel to the length of the solenoid. Its magnitude is

$$B = \mu_0 n I_{sol} = \left(4\pi \times 10^{-7} \text{ T·m/A}\right)\left[\left(30 \ \frac{\text{turns}}{\text{cm}}\right)\left(\frac{100 \text{ cm}}{1 \text{ m}}\right)\right](15.0 \text{ A}) = 5.65 \times 10^{-2} \text{ T}$$

Each side of the loop has a length $\ell = 2.00 \text{ cm} = 0.020\,0 \text{ m}$ and carries current $I_{loop} = 0.200 \text{ A}$. Since the plane of the loop is perpendicular to the magnetic field, the angle between either of the conducting sides of the loop and the magnetic field is $\theta = 90.0°$. Therefore, the magnitude of the magnetic force acting on each side of the loop is

$$F = BI_{loop}\,\ell(\sin 90.0°) = \left(5.65 \times 10^{-2} \text{ T}\right)(0.200 \text{ A})(0.020\,0 \text{ m})(1.00) = 2.26 \times 10^{-4} \text{ N} \qquad \lozenge$$

The drawing at the right shows the square loop with the magnetic field directed perpendicular to the plane of the loop. Since the magnetic force exerted on a current carrying conductor is perpendicular to both the magnetic field and the direction of the current, the force on each side of the loop lies in the plane of the loop and is perpendicular to that side.

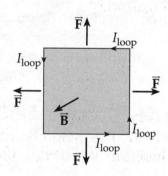

When the current flows counterclockwise around the loop, right-hand rule #1 shows that the forces are all directed away from the interior of the loop as indicated. Thus, the forces acting on the loop when its plane is perpendicular to the field tend to stretch the loop. They do not tend to rotate the loop about any axis, so the net torque acting on the loop is

$$\tau = 0 \qquad \lozenge$$

61. Two long parallel conductors carry currents $I_1 = 3.00$ A and $I_2 = 3.00$ A, both directed into the page in Figure P19.61. Determine the magnitude and direction of the resultant magnetic field at P.

Figure P19.61 (modified)

Solution

First, observe that the triangle shown in dashed lines is a right triangle, and the two marked angles are

$$\alpha = \sin^{-1}\left(\frac{12.0 \text{ cm}}{13.0 \text{ cm}}\right) = 67.4°, \text{ and } \beta = 90.0° - \alpha = 22.6°$$

At point P, the magnetic field contributed by wire 1 has a magnitude

$$B_1 = \frac{\mu_0 I_1}{2\pi r_1} = \frac{\left(4\pi \times 10^{-7} \text{ T·m/A}\right)(3.00 \text{ A})}{2\pi\left(5.00 \times 10^{-2} \text{ m}\right)} = 1.20 \times 10^{-5} \text{ T} = 12.0 \text{ } \mu\text{T}$$

This field is tangent to the circle centered on wire 1 and passing through P. Use of right-hand rule #2 shows it to be directed from P toward wire 2 along this tangent line. Thus, $\vec{B}_1$ is directed to the left and at 67.4° below the horizontal.

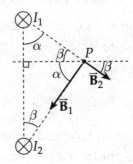

The contribution of wire 2 to the magnetic field at point P has magnitude

$$B_2 = \frac{\mu_0 I_2}{2\pi r_2} = \frac{\left(4\pi \times 10^{-7} \text{ T·m/A}\right)(3.00 \text{ A})}{2\pi\left(12.0 \times 10^{-2} \text{ m}\right)} = 5.00 \times 10^{-6} \text{ T} = 5.00 \text{ } \mu\text{T}$$

It is tangent to the circle centered on wire 2 and passing through P. From right-hand rule #2, this field contribution is directed away from wire 1 along this tangent line. Therefore, it is directed to the right and 22.6° below the horizontal.

The horizontal (x) and vertical (y) components of these field contributions are:

$$B_{1x} = -B_1 \cos 67.4° = -4.62 \text{ } \mu\text{T} \qquad B_{1y} = -B_1 \sin 67.4° = -11.1 \text{ } \mu\text{T}$$

$$B_{2x} = B_2 \cos 22.6° = +4.62 \text{ } \mu\text{T} \qquad B_{2y} = -B_2 \sin 22.6° = -1.92 \text{ } \mu\text{T}$$

The components of the resultant magnetic field at point P are:

$$B_x = B_{1x} + B_{2x} = 0 \qquad \text{and} \qquad B_y = B_{1y} + B_{2y} = -13.0 \text{ } \mu\text{T}$$

The resultant field at P is then

$$\vec{B} = 13.0 \text{ } \mu\text{T} \text{ directed toward the bottom of the page}$$

◊

65.   Protons having a kinetic energy of 5.00 MeV are moving in the positive x-direction and enter a magnetic field of 0.050 0 T in the z-direction, out of the plane of the page, and extending from $x = 0$ to $x = 1.00$ m as in Figure P19.65. (a) Calculate the y-component of the protons' momentum as they leave the magnetic field. (b) Find the angle α between the initial velocity vector of the proton beam and the velocity vector after the beam emerges from the field. [*Hint:* Neglect relativistic effects and note that $1 \text{ eV} = 1.60 \times 10^{-19} \text{ J}$.]

Solution

Note that we solve part (b) before solving part (a).

For a non-relativistic particle, the kinetic energy is $KE = \frac{1}{2}mv^2$, and the magnitude of the momentum may be expressed as $p = mv = \sqrt{2m(KE)}$.

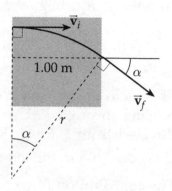

(b)   While the particle is in the magnetic field, it is moving perpendicular to the field and follows a circular arc. The magnetic force supplies the centripetal acceleration, so we determine the radius of the circular arc as follows:

$$qvB\sin 90° = m\frac{v^2}{r} \qquad \text{and} \qquad r = \frac{mv}{qB} = \frac{\sqrt{2m(KE)}}{qB}$$

or $r = \dfrac{\sqrt{2\left(1.67 \times 10^{-27} \text{ kg}\right)\left(5.00 \times 10^6 \text{ eV}\right)\left(1.60 \times 10^{-19} \text{ J/eV}\right)}}{\left(1.60 \times 10^{-19} \text{ C}\right)\left(0.050\,0 \text{ T}\right)} = 6.46 \text{ m}$

Then, considering the sketch given above, observe that the angle between the directions of the initial and final velocities of the protons is

$$\alpha = \sin^{-1}\left(\frac{1.00 \text{ m}}{r}\right) = \sin^{-1}\left(\frac{1.00 \text{ m}}{6.46 \text{ m}}\right) = 8.90° \qquad\qquad ◊$$

(a)   Since the magnetic force is always perpendicular to the motion of the protons, it does no work on them, and they maintain constant speed while in the field. Thus, $v_f = v_i = \sqrt{2(KE)/m}$. As the protons emerge from the field, the y-component of their momentum is

$$\left(p_f\right)_y = m\left(v_f\right)_y = -mv_f \sin\alpha = -\sin\alpha\sqrt{2m(KE)}$$

or $\quad \left(p_f\right)_y = -\sin(8.90°)\sqrt{2\left(1.67 \times 10^{-27}\ \text{kg}\right)\left(5.00 \times 10^6\ \text{eV}\right)\left(1.60 \times 10^{-19}\ \text{J/eV}\right)}$

$\left(p_f\right)_y = -8.00 \times 10^{-21}\ \text{kg·m/s} \qquad\qquad\qquad\qquad \lozenge$

Chapter 20

Induced Voltages and Inductance

NOTES FROM SELECTED CHAPTER SECTIONS

20.1 Induced emf and Magnetic Flux

An electric current can be produced by a changing magnetic field. (A steady magnetic field cannot produce a current.) *It is customary to say that an induced emf is produced in a secondary circuit by the changing magnetic field in a primary circuit.* The emf is induced by a change in a quantity called the magnetic flux rather than simply by a change in the magnetic field.

As described In Chapter 19, magnetic field lines can be drawn to illustrate the direction and relative strength of a magnetic field throughout a region of space. Consider a plane area, A, immersed in a magnetic field with field lines intersecting the area; the magnetic flux is the product of the magnetic field, the area, and the cosine of the angle between the direction of the magnetic field and the perpendicular (normal) to the area. Magnetic flux, Φ_B is proportional to the number of magnetic field lines passing through the area and has SI units of webers (1 Wb = 1 T·m^2).

20.2 Faraday's Law of Induction

The emf induced in a circuit (loop) is proportional to the time rate of change of magnetic flux through the circuit; the resulting current is called an induced current. The polarity of the induced emf can be predicted by **Lenz's law**:

> **The polarity of an induced emf is such that it produces a current whose magnetic field opposes the change in magnetic flux through the loop. That is the induced current will be in the direction that will tend to maintain the original flux through the loop.**

20.3   Motional emf

A motional emf (potential difference) will be maintained across a conductor moving in a magnetic field as long as the direction of motion through the field is not parallel to the field direction. If the motion is reversed, the polarity of the potential difference will also be reversed.

20.5   Generators

An **alternating current (AC) generator** consists of a wire loop rotated by some external means in a magnetic field. The ends of the loop are connected to **slip rings** that rotate with the loop; connections to the external circuit are made by stationary brushes in contact with the slip rings. The induced emf varies sinusoidally with time with a frequency of 60 Hz (in the US and Canada.)

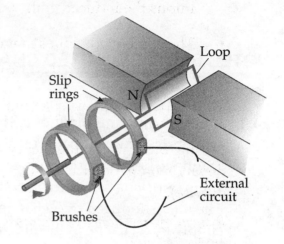

In a **direct current (DC) generator** the contacts to the rotating loop are made by a split ring, or **commutator**. In this design, the output voltage always has the same polarity (always in the same direction) and varies in magnitude from zero to some maximum value.

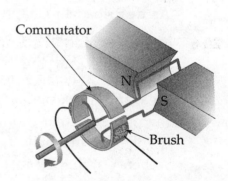

In the **operation of a motor,** a current is supplied to the loop by an external source of emf, and the magnetic torque on the current-carrying loop causes it to rotate thereby doing mechanical work.

20.6   Self-Inductance
20.7   *RL* Circuits

When the switch is closed in a circuit consisting of a resistor, coil, and a battery, the current does not instantly go from zero to its maximum value of $\mathcal{E}/R$. The current, increasing in the coil, gives rise to an emf which opposes the emf of the battery. This effect, called **self-induction**, can be considered in the following steps:

- Current, initially at zero, increases through the coil resulting in an increasing magnetic flux through the coil.

- The changing flux induces an emf in the coil (Faraday's law).

- The induced emf is proportional to the rate at which the current is changing.

- The induced emf opposes the emf of the battery (Lenz's law).

- The net potential difference across the resistor is the emf of the battery minus the induced emf.

- The current continues to increase to a maximum value of $\mathcal{E}/R$; as the current increases in value, the rate of increase becomes smaller.

The property of a coil, solenoid or other device (acting as an inductor) to limit the rate of change of current is called inductance; denoted by the symbol L and having SI units of the henry (H). The inductance of a device (an inductor) depends on geometric factors (cross-sectional area, length and number of turns of wire).

Remember, resistance (R) limits the current; inductance (L) limits the rate at which the current changes.

20.8 Energy Stored in a Magnetic Field

In an RL circuit, the rate at which energy is supplied by the battery equals the sum of the rate at which energy is dissipated in the resistor plus the rate at which energy is stored in the magnetic field of the inductor. The energy stored in the magnetic field of an inductor is proportional to the square of the current in that inductor.

EQUATIONS AND CONCEPTS

The **total magnetic flux** (Φ_B) through a plane area placed in a uniform magnetic field depends on the angle between the direction of the magnetic field and the direction perpendicular (normal) to the surface area.

$$\Phi_B \equiv B_\perp A = BA\cos\theta \qquad (20.1)$$

The **maximum flux** through the area occurs when the magnetic field is perpendicular to the surface area (along the direction of the normal). When $\theta = 90°$ the magnetic field is parallel to the plane of the surface area, and the flux through the area is zero. The unit of magnetic flux is the weber, $\mathrm{Wb} = 1\ \mathrm{T}\cdot\mathrm{m}^2$.

$$\Phi_{B,\max} = BA$$

normal to area

$\vec{B}$

Faraday's law of induction states that the average emf induced in a circuit during a time interval Δt is proportional to the rate of change of magnetic flux through the circuit. *The minus sign is included to indicate the polarity of the induced emf, which can be found by use of Lenz's law.*

$$\mathcal{E} = -N\frac{\Delta\Phi_B}{\Delta t} \qquad (20.2)$$

Lenz's law states that the polarity of the induced emf (and the direction of the associated current in a closed circuit) produces a current whose magnetic field opposes the change in the flux through the loop. *That is, the induced current tends to maintain the original flux through the circuit.* Note: If the circuit contains a source of emf (i.e. a battery) the current in the circuit may not be in the direction of the induced emf.

A **motional emf** is induced in a conductor of length, ℓ, moving with speed, v, perpendicular to a magnetic field. *Remember the significance of the minus sign as required by Lenz's law.*

$$|\mathcal{E}| = \frac{\Delta \Phi_B}{\Delta t} = B\ell v \qquad (20.4)$$

An **induced current** will be present when a moving conductor is part of a complete circuit of resistance, R. *Use the right-hand-rule and Lenz's law to confirm the direction of current as shown in the figure.*

$$I = \frac{|\mathcal{E}|}{R} = \frac{B\ell v}{R} \qquad (20.5)$$

A **sinusoidally varying emf** is produced when a loop of wire with N turns and cross-sectional area A rotates with an angular velocity ω (measured in radians per second) in a magnetic field. For a given loop, the maximum value of the induced emf will be proportional to the angular velocity of the loop.

$$\mathcal{E} = NBA\omega \sin \omega t \qquad (20.7)$$

$$\mathcal{E}_{max} = NBA\omega \qquad (20.8)$$

Inductance (L) is characteristic of a conducting device (e.g. coil, solenoid, toroid, coaxial cable). The inductance of a given device, for example a coil, depends on its physical makeup: diameter, number of turns, type of material (core) on which the wire is wound, and other geometric parameters. SI unit of inductance is the henry (H).

$$1\,H = 1\,V \cdot s/A$$

A **self-induced emf** will be produced in a circuit in which the current is changing. *The negative sign indicates that the induced emf opposes the <u>change</u> in the current.*

$$\mathcal{E} \equiv -L\frac{\Delta I}{\Delta t} \qquad (20.9)$$

The **inductance of a circuit element** (e.g. circular coil or solenoid) can be calculated if the magnetic flux is known for a given current (Eq. 20.10). The inductance can also be stated in terms of the geometry of the inductor. Equations 20.11a and 20.11b are used to calculate the inductance of a coil or solenoid.

$$L = \frac{N\Phi_B}{I} \tag{20.10}$$

$$L = \frac{\mu_0 N^2 A}{\ell} \tag{20.11a}$$

$$L = \mu_0 n^2 V \tag{20.11b}$$

$$n = N/\ell$$

$$V = A\ell = volume\ of\ solenoid$$

A series *RL* circuit is shown in the figure at right. The circuit elements are a battery, resistor, inductor and switch. *The inductor opposes any change in the current in the circuit.*

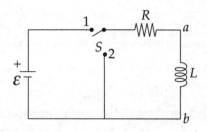

A **self-induced emf** ("back" emf) will be generated in a coil when the current in the coil is changing. The emf will be proportional to the rate of change of the current and has a polarity that opposes the *change* occurring in the current.

$$\mathcal{E}_L = -L\frac{\Delta I}{\Delta t} \tag{20.13}$$

If the switch in the series circuit above is closed (moved to position 1) at time $t = 0$, the current in the circuit will increase as shown in the figure at right, approaching a maximum value of $I_{max} = \mathcal{E}/R$.

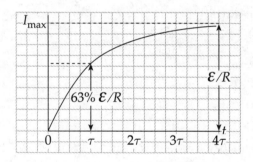

The rate at which current increases in a series *RL* circuit as a function of time is shown in the figure above and stated by this equation.

$$I = \frac{\mathcal{E}}{R}\left(1 - e^{-Rt/L}\right)$$

The **time constant of the circuit** (*τ*) is the time required for the current to reach 63 percent of its maximum value.

$$\tau \equiv \frac{L}{R} \tag{20.14}$$

The **energy stored in the magnetic field** of an inductor is proportional to the square of the current.

$$PE_L = \tfrac{1}{2}LI^2 \qquad (20.15)$$

The **energy stored in the electric field** of a charged capacitor is proportional to the square of the potential difference between the plates.

$$PE_C = \tfrac{1}{2}C(\Delta V)^2$$

SUGGESTIONS, SKILLS, AND STRATEGIES

Use the right-hand rule when applying Lenz's law to determine the direction of an induced current. **Remember, the induced current in a circuit will have a direction which tends to maintain the flux through the circuit.**

Consider two single-turn, concentric coils **lying in the plane of the paper** as shown in the figure. The outside coil is part of a circuit containing a resistor (R), a battery ($\mathcal{E}$) and switch (S). The inner coil is not part of the circuit. When the switch is moved from "**open**" to "**closed**" the direction of the induced current in the inner coil can be predicted by Lenz's law.

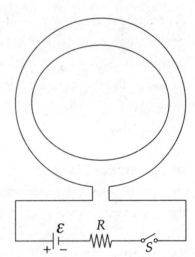

Consider the following steps:

(1) When the switch is in the "**open**" position as shown, there will be no current in the circuit.

(2) When the switch is moved to the "**closed**" position, there will be a clockwise current in the outside coil.

(3) Magnetic field lines due to current in the outside coil will be directed into the page through the area enclosed by the coil. **Use right-hand rule #1 to confirm this.** Magnetic flux into the page will penetrate the entire area enclosed by the outside coil **including the area of the inner coil.**

(4) By Faraday's law, the increasing flux produces an induced emf (and current) in the inner coil.

(5) Lenz's law requires that the induced current have a direction which will tend to maintain the initial flux condition **(which in this case was zero).** By using **right-hand rule #2,** you should be able to determine that the direction of the induced current in the inner coil must be counterclockwise, contributing to a flux out of the page. Try this!

As a second example you should follow steps similar to those above to predict the direction of the induced current in the inner coil when the switch is moved from "**closed**" to "**open**".

REVIEW CHECKLIST

▷ Calculate the emf (or current) induced in a circuit when the magnetic flux through the circuit is changing in time. The variation in flux might be due to a change in: (i) the area of the circuit, (ii) the magnitude of the magnetic field, (iii) the direction of the magnetic field, or (iv) the orientation/location of the circuit in the magnetic field.

▷ Apply Lenz's law to determine the direction of an induced emf or current. You should also understand that Lenz's law is a consequence of the law of conservation of energy.

▷ Calculate the emf induced between the ends of a conducting bar as it moves through a region where there is a constant magnetic field (motional emf).

▷ Define self-inductance, L, of a circuit in terms of appropriate circuit parameters. Calculate the total magnetic energy stored in a magnetic field if you are given the values of the inductance of the device with which the field is associated and the current in the circuit.

▷ Qualitatively describe the manner in which the instantaneous value of the current in an RL circuit changes while the current is either increasing or decreasing with time.

SOLUTIONS TO SELECTED END-OF-CHAPTER PROBLEMS

5.   A long, straight wire lies in the plane of a circular coil with a radius of 0.010 m. The wire carries a current of 2.0 A and is placed along a diameter of the coil. (a) What is the net flux through the coil? (b) If the wire passes through the center of the coil and is perpendicular to the plane of the coil, find the net flux through the coil.

Solution

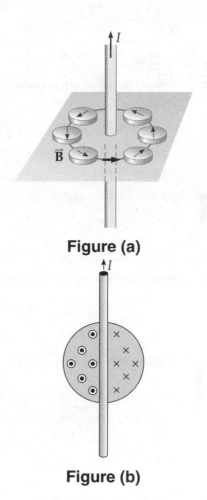

The magnetic field lines produced by a current in a long, straight wire are circular, centered on the wire, and in a plane perpendicular to the wire as shown in Figure (a). Note that if the current is toward the top of the page, the field lines go into the page on the right side of the wire and come out of the page on the left side of the wire.

Figure (a)

(a)   If the long, straight wire is along the diameter of a flat circular coil as shown in Figure (b), every magnetic field line passing through the area of the coil directed into the page on the right side of the long, straight wire reemerges from the page through the area of the coil on the left side of the wire. Thus, there are just as many lines passing through the area of the coil in one direction as there are passing through this area in the opposite direction, and the net flux through the area enclosed by the coil is **zero**.   ◊

Figure (b)

(b)   If the wire passes through the center of the coil and is perpendicular to the plane of the coil, then all of the field lines produced by the current in the long, straight wire lie in planes parallel to the plane of the coil. Thus, none of these lines pass through the area enclosed by the coil, and the flux through the coil is again **zero**.   ◊

9.   A square, single-turn coil 0.20 m on a side is placed with its plane perpendicular to a constant magnetic field. An emf of 18 mV is induced in the coil winding when the area of the coil decreases at the rate of 0.10 m²/s. What is the magnitude of the magnetic field?

Solution

When a flat coil is in a uniform magnetic field, the flux passing through the coil is given by

$$\Phi_B = BA\cos\theta$$

where B is the magnitude of the magnetic field, A is the area enclosed by the coil, and θ is the angle between the normal line to the plane of the coil and the direction of the magnetic field. When the plane of the coil is perpendicular to the field, the normal line is parallel to the field and $\theta = 0°$. Thus, the flux through the coil reduces to $\Phi_B = BA$.

If the field is constant in time while the area enclosed by the coil is changing, this flux is changing at a rate of

$$\frac{\Delta\Phi_B}{\Delta t} = B\left(\frac{\Delta A}{\Delta t}\right)$$

and the magnitude of the emf induced in the coil will be

$$|\mathcal{E}| = \left|\frac{\Delta\Phi_B}{\Delta t}\right| = B\left|\frac{\Delta A}{\Delta t}\right|$$

If $\mathcal{E} = 18$ mV when $\Delta A/\Delta t = -0.10$ m²/s, the magnetic field magnitude must be

$$B = \frac{|\mathcal{E}|}{|\Delta A/\Delta t|} = \frac{18\times10^{-3}\ \text{V}}{0.10\ \text{m}^2/\text{s}} = 0.18\ \text{T} \qquad \Diamond$$

16. A circular coil enclosing an area of 100 cm² is made of 200 turns of copper wire. The wire making up the coil has resistance of 5.0 Ω, and the ends of the wire are connected to form a closed circuit. Initially, a 1.1-T uniform magnetic field points perpendicularly upward through the plane of the coil. The direction of the field then reverses so that the final magnetic field has a magnitude of 1.1 T and points downward through the coil. If the time required for the field to reverse directions is 0.10 s, what average current flows through the coil during that time?

Solution

The magnitude of the average induced emf in a coil of fixed size is given by

$$|\mathcal{E}_{av}| = N \frac{|\Delta \Phi_B|}{\Delta t} = N \frac{|\Delta(B \cos \theta)| A}{\Delta t}$$

Here, N is the number of turns on the coil, B is the magnitude of the magnetic field, A is the area enclosed by the coil, and θ is the angle between the line normal to the plane of the coil and the direction of the magnetic field.

In the given situation, the field initially has a magnitude of 1.1 T and is directed parallel $(\theta = 0°)$ to the normal line. After an interval of $\Delta t = 0.10$ s, the field has a magnitude of 1.1 T directed in the opposite $(\theta = 180°)$ direction. Thus,

$$|\mathcal{E}_{av}| = (200) \frac{|(1.1 \text{ T}) \cos 180° - (1.1 \text{ T}) \cos 0°| (100 \text{ cm}^2)}{0.10 \text{ s}} \left(\frac{1 \text{ m}}{10^2 \text{ cm}} \right)^2 = 44 \text{ V}$$

and the average current in the coil during this interval is

$$I_{av} = \frac{|\mathcal{E}_{av}|}{R} = \frac{44 \text{ V}}{5.0 \text{ Ω}} = 8.8 \text{ A}$$

◊

21. An automobile has a vertical radio antenna 1.20 m long. The automobile travels at 65.0 km/h on a horizontal road where Earth's magnetic field is 50.0 μT, directed toward the north and downwards at an angle of 65.0° below the horizontal. (a) Specify the direction the automobile should move in order to generate the maximum motional emf in the antenna, with the top of the antenna positive relative to the bottom. (b) Calculate the magnitude of this induced emf.

Solution

We want free electrons within the moving antenna to experience a downward force exerted on them by Earth's magnetic field. Since the magnetic force experienced by moving charges is perpendicular to the field exerting that force, only the horizontal component, B_h, of the Earth's field is effective in exerting a downward force on these electrons. The magnitude of this force is then

$$F = qvB_h \sin\theta \text{ where } B_h = (50.0 \ \mu\text{T})\cos 65.0° = 21.1 \ \mu\text{T}$$

is the northward directed horizontal component of Earth's field. Here, v is the magnitude of the antenna's velocity, and θ is the angle this velocity makes with B_h (that is, with the northward direction).

Electrons will migrate to the bottom of the antenna until an opposing force exerted by the developing electric field within the antenna has a magnitude equal to that of the magnetic force. When this occurs, $qE = qvB_h \sin\theta$ or $E = vB_h \sin\theta$, and the potential difference between the ends of an antenna of length ℓ is

$$\mathcal{E} = E\ell = v\ell B_h \sin\theta$$

(a) If the emf discussed above is to have maximum magnitude, it is necessary that $\sin\theta = 1.00$ or $\theta = 90.0°$. Thus, the automobile must move along an east-west line. If the magnetic force on negative charges is to be downwards (and hence upward for positive charges), right-hand rule #1 shows that the automobile must move **eastward** along this line. ◊

(b) With $\theta = 90.0°$, the magnitude of the induced emf is

$$\mathcal{E} = v\ell B_h = \left[(65.0 \text{ km/h})\left(\frac{0.278 \text{ m/s}}{1 \text{ km/h}}\right)\right](1.20 \text{ m})(21.1\times 10^{-6} \text{ T})$$

or $\mathcal{E} = 4.58\times 10^{-4}$ V $= 0.458$ mV ◊

29. Find the direction of the current in resistor R in Figure P20.29 after each of the following steps (taken in the order given). (a) The switch is closed. (b) The variable resistance in series with the battery is decreased. (c) The circuit containing resistor R is moved to the left. (d) The switch is opened.

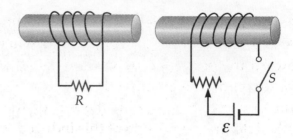

Figure P20.29

Solution

We shall refer to the coil on the right in Figure P20.29 as the primary coil and that on the left as the secondary coil. When switch S is closed, a current will flow clockwise (as viewed from the right end) around the loops of the primary coil. This current will produce a magnetic field that is <u>directed from the right to left along the axis of both the primary and the secondary coils</u>.

(a)   Immediately after switch S is closed, the current in the primary coil is increasing and the field produced by this current is increasing in magnitude. Thus, there is an increasing magnetic flux directed from right to left through the loops of the secondary coil.

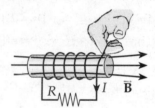

Figure 1

An induced current will flow in the secondary coil in a manner that produces a magnetic field directed left to right along the axis of the secondary coil, opposing the increasing flux from the primary coil. We use right-hand-rule #2 to determine the direction of the current in the resistance R. When one of the loops of the secondary coil is gripped with the right hand so the fingers point to the right inside the coil (in the direction of the field produced by the induced current), the thumb points in the direction the induced current must flow. Note in Figure 1 that this means the current must be directed out of the page over the top of the secondary coil and flow **right to left** through the resistance R. ◊

(b)   If, while switch S is closed, the variable resistance in the primary circuit is decreased, the current in the primary coil increases producing an increasing flux directed right to left through the secondary coil. As above, the induced current in the secondary coil will be directed out of the page over the top of the secondary coil and flow **right to left** through R so as to produce a magnetic field toward the right opposing the increasing flux from the primary coil. ◊

(c)   If the secondary coil is now moved toward the left, the
magnitude of the magnetic field directed right to left
along its axis by the primary coil will be decreasing.
Hence, the secondary coil will have a decreasing flux
directed toward the left through its loops.

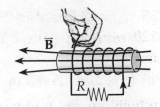

Figure 2

This induces a current that opposes the decrease in the flux from the primary
coil by serving as an additional source of flux directed toward the left through
the secondary coil. Using right-hand-rule #2, grip one of the loops of the
secondary coil with the right hand so the fingers point toward the left along
the axis, in the direction of the field produced by the induced current. As
Figure 2 shows, the induced current flows into the page over the top of the
secondary coil, and hence, **left to right** through R.   ◊

(d)   When switch S is opened, the current in the primary coil must cease to flow.
As this current is decreasing, the primary coil is producing a decreasing flux
directed toward the left in the secondary coil. To oppose this decrease in the
flux to the left, the induced current in the secondary coil must produce
additional flux directed toward the left through the loops of the secondary coil.
As above, Figure 2 and right-hand-rule #2 show that the induced current will
be directed **left to right** through R in this case.   ◊

35. In a model AC generator, a 500-turn rectangular coil 8.0 cm by 20 cm rotates at 120 rev/min in a uniform magnetic field of 0.60 T. (a) What is the maximum emf induced in the coil? (b) What is the instantaneous value of the emf in the coil at $t = (\pi/32)$ s? Assume that the emf is zero at $t = 0$. (c) What is the smallest value of t for which the emf will have its maximum value?

Solution

When a coil rotates in a uniform magnetic field, the emf induced in that coil is given by $\mathcal{E} = NBA\omega\sin(\omega t)$. Here, N is the number of turns on the coil, B is the magnitude of the magnetic field, A is the area enclosed by the coil, and ω is the angular frequency of the coil.

(a)   The angular frequency of the coil is

$$\omega = 120 \ \frac{\text{rev}}{\text{min}}\left(\frac{2\pi \ \text{rad}}{1 \ \text{rev}}\right)\left(\frac{1 \ \text{min}}{60 \ \text{s}}\right) = 4\pi \ \text{rad/s}$$

The maximum emf occurs at times when $\sin\omega t = 1.0$ and has a value of

$$\mathcal{E}_{max} = NBA\omega = (500)(0.60 \ \text{T})\big[(0.080 \ \text{m})(0.20 \ \text{m})\big](4\pi \ \text{rad/s}) = 60 \ \text{V} \qquad \Diamond$$

(b)   At $t = (\pi/32)$ s, the instantaneous emf is

$$\mathcal{E} = \mathcal{E}_{max}\sin\omega t = (60 \ \text{V})\sin\left[\left(4\pi \ \frac{\text{rad}}{\text{s}}\right)\left(\frac{\pi}{32} \ \text{s}\right)\right] = 57 \ \text{V} \qquad \Diamond$$

Note that your calculator must be set to operate in radians mode (rather than degrees) to do the above calculation correctly.

(c)   The first time $\sin\omega t$ (and hence $\mathcal{E} = \mathcal{E}_{max}\sin\omega t$) will achieve a maximum value is when

$$\omega t = \frac{\pi}{2} \ \text{rad} \qquad\qquad \text{or} \qquad\qquad t = \frac{\pi \ \text{rad}}{2\omega} = \frac{\pi \ \text{rad}}{2(4\pi \ \text{rad/s})} = 0.13 \ \text{s} \qquad \Diamond$$

39.   A solenoid of radius 2.5 cm has 400 turns and a length of 20 cm. Find (a) its inductance and (b) the rate at which current must change through it to produce an emf of 75 mV.

Solution

(a)   The self-inductance of a solenoid is given by $L = \mu_0 N^2 A/\ell$ where N is the number of turns on the solenoid, ℓ is the length of the solenoid, and A is its cross-sectional area. For the specified solenoid,

$$L = \frac{\left(4\pi \times 10^{-7}\ \text{T·m/A}\right)\left(400\right)^2 \left[\pi \left(2.5 \times 10^{-2}\ \text{m}\right)^2\right]}{0.20\ \text{m}} = 2.0 \times 10^{-3}\ \text{H} = 2.0\ \text{mH} \qquad \Diamond$$

(b)   The magnitude of the emf induced in an element having self-inductance L is $\left|\mathcal{E}\right| = L(\Delta I/\Delta t)$. Thus, the required rate at which the current must change to induce an emf of 75 mV in the specified solenoid is

$$\frac{\Delta I}{\Delta t} = \frac{\left|\mathcal{E}\right|}{L} = \frac{75 \times 10^{-3}\ \text{V}}{2.0 \times 10^{-3}\ \text{H}} = 38\ \text{A/s} \qquad \Diamond$$

45. Calculate the resistance in an RL circuit in which $L = 2.50$ H and the current increases to 90.0% of its final value in 3.00 s.

Solution

When the switch in a series RL circuit is closed at time $t = 0$, the current in the circuit is given as a function of time by

$$I = \frac{\mathcal{E}}{R}\left(1 - e^{-t/\tau}\right) = I_{max}\left(1 - e^{-t/\tau}\right)$$

where $\tau = \dfrac{L}{R}$

Thus, if $I = 0.900\, I_{max}$ at $t = 3.00$ s,

we must have $0.900 = 1 - e^{-(3.00\text{ s})/\tau}$

or $e^{-(3.00\text{ s})/\tau} = 0.100$

Taking the natural logarithm of both sides of the last result yields

$$-\frac{3.00\text{ s}}{\tau} = \ln\left(0.100\right)$$

or $$\tau = \frac{-3.00\text{ s}}{\ln(0.100)} = 1.30\text{ s}$$

We then find $$R = \frac{L}{\tau} = \frac{2.50\text{ H}}{1.30\text{ s}} = 1.92\ \Omega \qquad \Diamond$$

49.   A 24-V battery is connected in series with a resistor and an inductor, with $R = 8.0\ \Omega$ and $L = 4.0$ H, respectively. Find the energy stored in the inductor (a) when the current reaches its maximum value and (b) one time constant after the switch is closed.

Solution

If a coil has self-inductance L, an energy input is required to establish a current in it. The work required to overcome the opposing induced emf and establish the current is $PE_L = \frac{1}{2}LI^2$. This is labeled potential energy because it is useful to think of it as stored in the magnetic field surrounding the inductor when current I flows. The inductor releases this energy attempting to prevent the current from decreasing when the power source is removed or turned down.

(a)   After the circuit has been intact for a very long time (as $t \to \infty$), the current reaches its final value of $I_{max} = \mathcal{E}/R = 24\ \text{V}/8.0\ \Omega = 3.0$ A. The stored energy at this time is

$$PE_L = \tfrac{1}{2}LI_{max}^2 = \tfrac{1}{2}(4.0\ \text{H})(3.0\ \text{A})^2 = 18\ \text{J} \qquad \Diamond$$

(b)   When one time constant has elapsed since the circuit was completed (that is, at $t = \tau = L/R$) the current is 63.2% of the final value. Since the final current is $I_{max} = 3.0$ A, the current at $t = \tau$ is $I = 0.632(3.0\ \text{A}) = 1.9$ A, and the stored energy is

$$PE_L = \tfrac{1}{2}LI^2 = \tfrac{1}{2}(4.0\ \text{H})(1.9\ \text{A})^2 = 7.2\ \text{J} \qquad \Diamond$$

55. The plane of a square loop of wire with edge length $a = 0.200$ m is perpendicular to the Earth's magnetic field at a point where $B = 15.0$ μT, as in Figure P20.55. The total resistance of the loop and the wires connecting it to the ammeter is 0.500 Ω. If the loop is suddenly collapsed by horizontal forces as shown, what total charge passes through the ammeter?

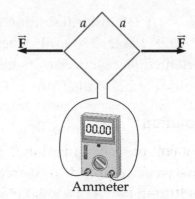

Figure P20.55

Solution

The average current in a conducting path may be written as $I_{av} = \Delta Q / \Delta t$ where ΔQ is the amount of charge that moves past a fixed point in the path in time Δt. Thus, the total charge passing through the ammeter in Figure P20.55 is

$$\Delta Q = I_{av}(\Delta t)$$

where I_{av} is the average induced current in the circuit as the area enclosed by the loop goes to zero, and Δt is the time required for the loop to collapse.

Since the magnitude and direction of the magnetic field is constant as the area of the loop is collapsing, the magnitude of the average induced emf is

$$\left|\mathcal{E}_{av}\right| = \frac{\left|\Delta\Phi_B\right|}{\Delta t} = \frac{B\left|\Delta A\right|}{\Delta t} = \frac{\left(15.0\times10^{-6}\text{ T}\right)\left|0-(0.200\text{ m})^2\right|}{\Delta t} = \frac{6.00\times10^{-7}\text{ T}\cdot\text{m}^2}{\Delta t}$$

If the resistance of the conducting path is $R = 0.500$ Ω, the average induced current in the circuit is

$$I_{av} = \frac{\left|\mathcal{E}_{av}\right|}{R} = \frac{\left(6.00\times10^{-7}\text{ T}\cdot\text{m}^2\right)/\Delta t}{0.500\ \Omega} = \frac{1.2\times10^{-6}\text{ C}}{\Delta t} = \frac{1.2\ \mu\text{C}}{\Delta t}$$

Thus, the total charge passing through the ammeter as the loop collapses is

$$\Delta Q = I_{av}(\Delta t) = \left(\frac{1.2\ \mu\text{C}}{\Delta t}\right)(\Delta t) = 1.2\ \mu\text{C} \qquad \Diamond$$

Note that the elapsed time Δt cancels in this calculation. This means that the same charge passes through the ammeter regardless of the time the loop takes to collapse.

59. The bolt of lightning depicted in Figure P20.59 passes 200 m from a 100-turn coil oriented as shown. If the current in the lightning bolt falls from 6.02×10^6 A to zero in 10.5 μs, what is the average voltage induced in the coil? Assume that the distance to the center of the coil determines the average magnetic field at the coil's position. Treat the lightning bolt as a long, vertical wire.

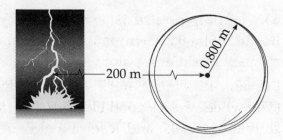

Figure P20.59

Solution

Treating the lightning bolt as a long vertical wire, the magnetic field lines produced by the current in the bolt will be horizontal circles centered on the bolt. Notice that this means the magnetic field is perpendicular to the plane of the 100-turn coil as it passes through the enclosed area of the coil. Also, because the diameter of the coil is small in comparison to the total distance to the lightning bolt, we will neglect the variation in magnitude of the magnetic field over the width of the coil.

When the current in the lightning bolt is I, the magnitude of the magnetic field produced at a distance of $r = 200$ m away is

$$B = \frac{\mu_0 I}{2\pi r} = \frac{\left(4\pi \times 10^{-7} \text{ T} \cdot \text{m/A}\right) I}{2\pi (200 \text{ m})} = \left(1.00 \times 10^{-9} \text{ T/A}\right) I$$

The average induced emf in the coil is then

$$\left|\mathcal{E}_{av}\right| = \frac{\left|\Delta \Phi_B\right|}{\Delta t} = \frac{\left|N(\Delta B) A\right|}{\Delta t} = \frac{100\left(1.00 \times 10^{-9} \text{ T/A}\right) |\Delta I| A}{\Delta t}$$

or

$$\left|\mathcal{E}_{av}\right| = \frac{100\left(1.00 \times 10^{-9} \text{ T/A}\right) \left|0 - 6.02 \times 10^6 \text{ A}\right| \pi (0.800 \text{ m})^2}{10.5 \times 10^{-6} \text{ s}} = 1.15 \times 10^5 \text{ V} = 115 \text{ kV} \qquad \Diamond$$

63. In Figure P20.63, the rolling axle, 1.50 m long, is pushed along horizontal rails at a constant speed $v = 3.00$ m/s. A resistor $R = 0.400\ \Omega$ is connected to the rails at points a and b, directly opposite each other. (The wheels make good electrical contact with the rails, so the axle, rails, and R form a closed-loop circuit. The only significant resistance in the circuit is R.)

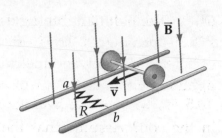

Figure P20.63

A uniform magnetic field $B = 0.800$ T is directed vertically downwards. (a) Find the induced current I in the resistor. (b) What horizontal force $\vec{F}$ is required to keep the axle rolling at constant speed? (c) Which end of the resistor, a or b, is at the higher electric potential? (d) After the axle rolls past the resistor, does the current in R reverse direction?

Solution

(a) As the rolling axle (of length $\ell = 1.50$ m) moves perpendicularly to the uniform magnetic field, an induced emf of magnitude $|\mathcal{E}| = B\ell v$ will exist between its ends. The current produced in the closed-loop circuit by this induced emf has magnitude

$$I = \frac{|\mathcal{E}|}{R} = \frac{B\,\ell v}{R} = \frac{(0.800\ \text{T})(1.50\ \text{m})(3.00\ \text{m/s})}{0.400\ \Omega} = 9.00\ \text{A} \qquad \Diamond$$

(b) The induced current flowing through the axle will cause the magnetic field to exert a retarding force of magnitude $F_r = BI\ell$ on the axle. The direction of this force will be opposite to that of the velocity $\vec{v}$ so as to oppose the motion of the axle. If the axle is to continue moving at constant speed, an applied force in the direction of $\vec{v}$ and having magnitude $F_{app} = F_r$ must be exerted on the axle.

$$F_{app} = BI\ell = (0.800\ \text{T})(9.00\ \text{A})(1.50\ \text{m}) = 10.8\ \text{N} \qquad \Diamond$$

(c) Using right-hand rule #1, observe that positive charges within the moving axle experience a magnetic force toward the rail containing point b, and negative charges experience a force directed toward the rail containing point a. Thus, the rail containing b will be positive relative to the other rail. Point b is then at a higher potential than a, and the conventional current goes from b to a through the resistor R. $\qquad \Diamond$

(d)   No. Both the velocity $\vec{v}$ of the rolling axle and the magnetic field $\vec{B}$ are unchanged. Thus, the polarity of the induced emf in the moving axle is unchanged, and the current continues to go from b to a through the resistor R.   ◊

Chapter 21

Alternating Current Circuits and Electromagnetic Waves

NOTES FROM SELECTED CHAPTER SECTIONS

21.1 Resistors in an AC Circuit

In an AC circuit both voltage and current vary sinsusoidally with time. If an alternating voltage is applied across a resistor, **the current in the resistor will be in phase with the voltage; current and voltage always vary in the same manner, and both reach their maximum values at the same time.** When stated in the form of Ohm's law, $\Delta V_{R,\max} = I_{\max}R$. The current is directed in opposite directions during successive half-cycles, and the average value of the current over one complete cycle is zero.

The value of the rms current in an AC circuit is equal to that of a direct current which would deliver the same internal energy to a resistor as does the rms current.

21.2 Capacitors in an AC Circuit

When an alternating voltage is applied across a capacitor, the voltage reaches its maximum value one quarter of a cycle after the current reaches its maximum value. *In this situation, it is common to say that the voltage across a capacitor always lags the current by 90°.*

A capacitor in an AC circuit acts to limit the current. The impeding effect of a capacitor is called **capacitative reactance, X_C.** When stated in the form of Ohm's law, $\Delta V_{L,\max} = I_{\max}X_L$. The value of the capacitative reactance is inversely proportional to the frequency of the applied voltage. Recall that when a dc voltage is applied to an RC circuit, the current goes to zero when the capacitor becomes fully charged.

21.3 Inductors in an AC Circuit

When an alternating voltage is applied across an inductor (coil), the voltage reaches its maximum value one quarter of a cycle before the current reaches its maximum value. *In this situation, we say that the voltage across an inductor always leads the current by 90°.* The extent by which an inductor impedes the current in an AC circuit is called **inductive reactance**, X_L. When stated in the form of Ohm's law, $\Delta V_{C,\text{max}} = I_{\text{max}} X_C$. This quantity is directly proportional to the frequency of the applied voltage. Recall that when a DC voltage is applied to an *RL* circuit, the inductor has no effect on the circuit after the current reaches its maximum (constant) value.

21.4   The *RLC* Series Circuit

At any given time the current in a series *RLC* (resistor, inductor and capacitor) circuit has the same amplitude and phase at all points in the circuit. The current through the individual elements reaches its maximum value at the same time. The voltages across the individual components (*R*, *L*, and *C*) do not reach their maximum values at the same time; they are out of phase with each other. The voltage across the resistor is in phase with the current, the voltage across the inductor leads the current by 90°, and the voltage across the capacitor lags the current by 90°.

The maximum applied voltage does not equal the sum of the individual maximum voltages:

$$\Delta V_{\text{max}} \neq \Delta V_{R,\text{max}} + \Delta V_{L,\text{max}} + \Delta V_{C,\text{max}}$$

The instantaneous applied voltage, however, does equal the sum of the instantaneous voltages across the individual components:

$$\Delta v = \Delta v_R + \Delta v_L + \Delta v_C$$

The phase relationships among the voltages in an AC circuit can be shown in a phasor diagram. The maximum voltage across each circuit component is represented by a rotating vector called a phasor.

The phase relationships stated above are shown in the figure to the right for the voltages across the resistor, inductor and capacitor in an AC circuit. The relative magnitudes of the voltages are indicated by the length of the corresponding phasor. **The phasor for the current, which lies along the +x-axis at *t*=0, is not shown in the diagram.**

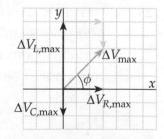

At any time, the instantaneous voltages (Δv_R, Δv_L, and Δv_C) across the circuit elements are the components of the maximum voltages ΔV_R, ΔV_L , and ΔV_C along the *y*-axis. Assume counterclockwise rotation of the phasors and confirm from the figure the following phase relationships:

Δv_R **is in phase with the current** (along the x-axis, but not shown in the figure).

Δv_L **leads the current by 90°** (reaches its maximum 90° before Δv_R achieves maximum).

Δv_C **lags the current by 90°** (reaches maximum 90° later than Δv_R).

The maximum voltage across the circuit, (ΔV_{max}), as shown in the diagram above, is found by adding ΔV_R, ΔV_L and ΔV_C as vector quantities. ΔV_{max} is shown as a rotating phasor in the figure on the right.

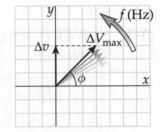

The instantaneous voltage across the circuit is the component of ΔV_{max} along the y-axis.

The **phase angle (ϕ)** indicates the degree by which the maximum voltage is out of phase with the current.

The **impedance, Z,** is a parameter which represents the combined effect of R, X_C and X_L in limiting current in an ac circuit. When stated in the form of Ohm's law, $\Delta V_{max} = I_{max} Z$. The value of Z depends on the values of R, L, C and the frequency, f, of the voltage source.

A **reactance triangle** showing R, X_L, X_C and Z can be drawn directly from the phasor diagram showing the voltages. This is possible because ΔV_R, ΔV_L and ΔV_C are each proportional to the current for given values of R, X_L, and X_C. Notice that Equation 21.13 follows directly from the reactance triangle.

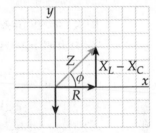

21.5 Power in an AC Circuit

The power delivered by the generator in an AC circuit is converted to internal energy in the resistor. **There is no power loss in an ideal inductor or an ideal capacitor.** The power delivered to the circuit is determined by the power factor, $\cos \phi$.

21.6 Resonance in a Series *RLC* Circuit

As the frequency of the source (generator) increases, the resistance (R) in the circuit remains constant; inductive reactance (X_L) increases, and capacitative reactance (X_C) decreases. **At the resonance frequency, f_0 (when $X_L = X_C$) the voltage across the capacitor is equal to but 180 degrees out of phase (oppositely directed) with the voltage across the inductor. In this case, $\Delta v_L + \Delta v_C = 0$, and** the impedance (Z) has its minimum value and is equal to R. *At resonance the current in the circuit is maximum, limited only by the value of R.*

21.7   The Transformer

A transformer is a device designed to increase or decrease an AC voltage. Energy conservation requires a corresponding decrease or increase in current. In an ideal transformer, the power input to the primary equals the power output at the secondary (load). In practice, output power is less than input power; this is due in part to eddy currents induced in the transformer core. In its simplest form, a transformer consists of a primary coil of N_1 turns and a secondary coil of N_2 turns, both wound onto a common soft iron core. A step-up transformer has $N_2 > N_1$ and in a step-down transformer $N_1 > N_2$.

21.8   Maxwell's Predictions
21.9   Hertz's Confirmation of Maxwell's Predictions

Maxwell's equations are the fundamental laws governing the behavior of electric and magnetic fields. The theory Maxwell developed is based upon the following four pieces of information:

- Electric field lines originate on positive charges and terminate on negative charges. The electric field due to a point charge can be determined at a location by applying Coulomb's force law to a positive test charge placed at that location.

- Magnetic field lines always form closed loops; that is, they do not begin or end at any point.

- A varying magnetic field induces an emf and hence, an electric field. This is a statement of Faraday's law (Chapter 20).

- Magnetic fields are generated by moving charges (or currents), as summarized in Ampère's law (Chapter 19).

21.11 Properties of Electromagnetic Waves

Following is a summary of the properties of electromagnetic waves:

- **Electromagnetic waves are transverse waves** and travel through empty space with the speed of light, $c = 1/\sqrt{\epsilon_0 \mu_0}$.

- **Radiated waves exist as electric and magnetic fields which oscillate perpendicular to each other.** The plane in which the oscillations occur is perpendicular to the direction of wave propagation.

- The ratio of $|\vec{E}|$ to $|\vec{B}|$ in empty space has a constant value, the speed of light in vacuum.

- Electromagnetic waves carry both energy and momentum.

21.12 The Spectrum of Electromagnetic Waves

All electromagnetic waves are produced by accelerating charges. Several types of electromagnetic waves are characterized by a "typical" range of frequencies or wavelengths. The following are listed in order of increasing frequency.

- **Radio waves** (~ 10 km$>\lambda>\sim 1$ cm) are the result of electric charges accelerating through a conducting wire (antenna).

- **Microwaves** (~ 1 cm$>\lambda>\sim 1$ mm) are generated by electronic devices.

- **Infrared waves** (~ 1 mm$>\lambda>\sim 700$ nm) are produced by high temperature objects and molecules.

- **Visible light** (~ 700 nm$>\lambda>\sim 400$ nm) is produced by the rearrangement of electrons in atoms and molecules.

- **Ultraviolet (UV) light** (~ 400 nm$>\lambda>\sim 1$ nm) is an important component of radiation from the Sun.

- **X-rays** (~ 10 nm$>\lambda>\sim 10^{-4}$ nm) are produced when high-energy electrons strike a target of high atomic number (e.g. metal or glass).

- **Gamma rays** ($\sim 10^{-1}$ nm$>\lambda>\sim 10^{-5}$ nm) are emitted by radioactive nuclei.

EQUATIONS AND CONCEPTS

A **series alternating current circuit** with a sinusoidal source of emf is shown in the figure to the right. The rectangle □ used here represents the circuit element(s) which, in a particular case, may be a resistor R, a capacitor C, an inductor L, or some combination of the above.

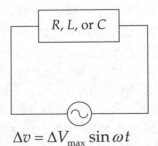

$$\Delta v = \Delta V_{max} \sin \omega t$$

The **output voltage of an AC generator** is sinusoidal where Δv is the instantaneous voltage and ΔV_{max} is the maximum voltage.

$$\Delta v = \Delta V_{max} \sin(2\pi ft) \qquad (21.1)$$

Root-mean-square (rms) values of current and voltage are those values to which measuring instruments usually respond. *You should notice that ΔV_{rms} and I_{rms} are in the same ratio as ΔV_{max} and I_{max}.*

$$I_{rms} = \frac{I_{max}}{\sqrt{2}} = 0.707\, I_{max} \qquad (21.2)$$

$$\Delta V_{rms} = \frac{\Delta V_{max}}{\sqrt{2}} = 0.707\, \Delta V_{max} \qquad (21.3)$$

The **capacitive reactance (X_c)** is the impeding effect of a capacitor on the current in an AC circuit.

$$X_C \equiv \frac{1}{2\pi fC} \qquad (21.5)$$

The **inductive reactance (X_L)** is the impeding effect of a coil in an AC circuit.

$$X_L \equiv 2\pi fL \qquad (21.8)$$

The **rms values of voltage and current** in each circuit element can be expressed by equations which have the form of Ohm's law. *Similar equations relate the maximum values of voltage and current.*

$$\Delta V_{R,rms} = I_{rms} R \qquad (21.4a)$$

$$\Delta V_{C,rms} = I_{rms} X_C \qquad (21.6)$$

$$\Delta V_{L,rms} = I_{rms} X_L \qquad (21.9)$$

The **maximum (or rms) voltage** across an *RLC* circuit can be found in terms of the respective voltage values across the individual components.

$$\Delta V_{max} = \sqrt{\Delta V_R{}^2 + (\Delta V_L - \Delta V_C)^2} \qquad (21.10)$$

$$\Delta V_{rms} = \sqrt{\Delta V_{R,rms}^2 + (\Delta V_{L,rms} - \Delta V_{C,rms})^2}$$

The maximum (or rms) circuit voltage can also be calculated in terms of the common circuit current and the values of resistance, inductive reactance, and capacitive reactance.

$$\Delta V_{max} = I_{max} \sqrt{R^2 + (X_L - X_C)^2} \qquad (21.12)$$

$$\Delta V_{rms} = I_{rms} \sqrt{R^2 + (X_L - X_C)^2}$$

Impedance (Z) is a parameter of the circuit defined by Eq. 21.13. Impedance is related to R, X_L and X_C as shown in the reactance triangle. It is possible to relate the maximum (or rms) voltage and maximum (or rms) current in the form of a generalized Ohm's law. *The SI unit of impedance is the Ohm.*

$$Z \equiv \sqrt{R^2 + (X_L - X_C)^2} \qquad (21.13)$$

The **phase angle** (the degree by which the instantaneous current and instantaneous voltage are out-of-step) can be determined from the reactance triangle shown above or from the voltage triangle. Refer to the phasor diagram in Section 21.4.

$$\tan \phi = \frac{X_L - X_C}{R} \qquad (21.15)$$

$$\tan \phi = \frac{\Delta V_L - \Delta V_C}{\Delta V_R} \qquad (21.11)$$

A **generalized form of Ohm's law** relates the maximum (or rms) voltage and maximum (or rms) current.

$$\Delta V_{max} = I_{max} Z \qquad (21.14)$$

$$\Delta V_{rms} = I_{rms} Z$$

The **average power** delivered by a generator (source of emf) to an *RLC* series circuit is directly proportional to cos ϕ, the power factor of the circuit. *There is zero power loss in ideal inductors and capacitors; the average power delivered by the source is converted to internal energy in the resistor.*

$$\mathcal{P}_{av} = I_{rms}^2 R \qquad (21.16)$$

$$\mathcal{P}_{av} = I_{rms} \Delta V_{rms} \cos \phi \qquad (21.17)$$

$$\cos \phi = \text{power factor}$$

The **resonance frequency**, f_0 is that frequency for which $X_L = X_C$ (and $Z = R$). *At this frequency the current has its maximum value and is in phase with the applied voltage.*

$$f_0 = \frac{1}{2\pi\sqrt{LC}} \qquad (21.19)$$

A **transformer** consists of a primary coil of N_1 turns and a secondary coil of N_2 turns wound on a common core. *In a step-up transformer, N_2 is greater than N_1.*

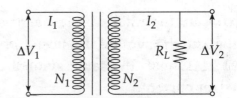

ΔV_1 = voltage across the primary
ΔV_2 = voltage across the secondary
I_1 = current in the primary
I_2 = current in the secondary

In **an ideal transformer**, the ratio of voltages is equal to the ratio of turns, and the ratio of currents is equal to the inverse of the ratio of turns. *Energy conservation requires a decrease in current to accompany an increase in voltage.*

$$\Delta V_2 = \frac{N_2}{N_1} \Delta V_1 \qquad (21.22)$$

$$I_1 \Delta V_1 = I_2 \Delta V_2 \qquad (21.23)$$

The **speed of an electromagnetic wave** is related to the permeability and permittivity of the medium through which it travels. In a vacuum the ratio of the electric to the magnetic field in an electromagnetic wave is constant and equal to the speed of light.

$$c = \frac{1}{\sqrt{\mu_0 \, \epsilon_0}} \tag{21.24}$$

$$\frac{E}{B} = c \tag{21.26}$$

The wave equation: The product of frequency and wavelength of an electromagnetic wave propagating in vacuum is constant and equal to c.

$$c = f\lambda \tag{21.31}$$

$$c = 2.997\,92 \times 10^8 \text{ m/s} \tag{21.25}$$

Permeability constant of free space.

$$\mu_0 = 4\pi \times 10^{-7} \text{ T·m/A}$$

Permittivity of free space.

$$\epsilon_0 = 8.854\,19 \times 10^{-12} \text{ C}^2/\text{N·m}^2$$

The **intensity (power per unit area) of electromagnetic waves** can be expressed in several alternate forms involving the maximum values of the electric and magnetic fields. *The energy carried by an electromagnetic wave is shared equally by the electric and magnetic fields.*

$$I = \frac{E_{max}B_{max}}{2\mu_0} \tag{21.27}$$

$$I = \frac{E_{max}^2}{2\mu_0 c} = \frac{c}{2\mu_0}B_{max}^2 \tag{21.28}$$

where $I = \dfrac{\mathcal{P}_{av}}{A} = $ Average power per unit area

Do not confuse the symbol, I, as used in Eqs. 21.27 and 21.28 with the same symbol used elsewhere for current.

The **magnitude of the momentum** delivered to a surface by an electromagnetic wave depends on the fraction of energy absorbed. *Radiation pressure is exerted on a surface as a result of momentum transfer by an electromagnetic wave.*

$$p = \frac{U}{c} \text{ (complete absorption)} \tag{21.29}$$

$$p = \frac{2U}{c} \text{ (complete reflection)} \tag{21.30}$$

SUGGESTIONS, SKILLS, AND STRATEGIES

PROCEDURE FOR SOLVING ALTERNATING CURRENT PROBLEMS

- First calculate as many of the unknown quantities such as X_L and X_C as possible. Be careful to use correct units; when calculating X_C, express capacitance in farads (not microfarads).

- Apply the general equation $\Delta V = IZ$ to the portion of the circuit that is of interest. That is, if you want to know the voltage drop across the combination of an inductor and a resistor, the equation reduces to $\Delta V_{max} = I_{max}\sqrt{R^2 + X_L^2}$ or $\Delta V_{rms} = I_{rms}\sqrt{R^2 + X_L^2}$.

REVIEW CHECKLIST

▷ Given an *RLC* series circuit in which values of resistance, inductance, capacitance, and the characteristics of the generator (source of emf) are known, calculate: (i) the rms voltage drop across each component, (ii) the rms current in the circuit, (iii) the phase angle by which the current leads or lags the voltage, (iv) the power expended in the circuit, and (v) the resonance frequency of the circuit.

▷ Understand the important function of transformers in the process of transmitting electrical power over large distances. Make calculations of primary to secondary voltage and current ratios for an ideal transformer.

▷ Be aware of the important pieces of information on which Maxwell based his theory of electromagnetic waves. Summarize the properties of electromagnetic waves. Describe the relative orientation of the magnetic field, electric field, and direction of propagation for a plane electromagnetic wave.

▷ Place the various types of electromagnetic waves in the correct sequence in the electromagnetic spectrum. Be aware that all forms of radiation are produced by accelerated charges and know the basis of production particular to each of the wave types.

SOLUTIONS TO SELECTED END-OF-CHAPTER PROBLEMS

5. An audio amplifier, represented by the AC source and the resistor R in Figure P21.5, delivers alternating voltages at audio frequencies to the speaker. If the source puts out an alternating voltage of 15.0 V (rms), the resistance R is 8.20 Ω, and the speaker is equivalent to a resistance of 10.4 Ω, what is the time-averaged power delivered to the speaker?

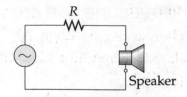

Figure P21.5

Solution

Observe that the resistance R and the resistance of the speaker $R_{speaker}$ are in series in this purely resistive AC circuit. Thus, the total resistance of the circuit is

$$R_{total} = R + R_{speaker} = 8.20\ \Omega + 10.4\ \Omega = 18.6\ \Omega$$

The rms current in this single branch circuit is then

$$I_{rms} = \frac{\Delta V_{rms,\ source}}{R_{total}} = \frac{15.0\ \text{V}}{18.6\ \Omega} = 0.806\ \text{A}$$

and the time-averaged power delivered to the speaker is

$$\mathcal{P}_{speaker} = I_{rms}^2 R_{speaker} = (0.806\ \text{A})^2 (10.4\ \Omega) = 6.76\ \text{W} \qquad \Diamond$$

9.   When a 4.0-μF capacitor is connected to a generator whose rms output is 30 V, the current in the circuit is observed to be 0.30 A. What is the frequency of the source?

Solution

When an AC voltage ΔV_C is applied to a capacitance C, the voltage and current are related by an Ohm's law-like expression, $\Delta V_C = IX_C$, where the capacitive reactance is given by

$$X_C = \frac{1}{2\pi fC} = \frac{1}{\omega C}$$

One may use either the maximum values for the voltage and current or the rms values for voltage and current since both $\Delta V_{C,\,max} = I_{max} X_C$ and $\Delta V_{C,\,rms} = I_{rms} X_C$ are valid expressions.

If $I_{rms} = 0.30$ A when $\Delta V_{C,\,rms} = 30$ V and $C = 4.0\ \mu$F, the capacitive reactance of the capacitor is

$$X_C = \frac{\Delta V_{C,\,rms}}{I_{rms}} = \frac{30\text{ V}}{0.30\text{ A}} = 100\ \Omega$$

and the frequency of the AC voltage source must be

$$f = \frac{1}{2\pi X_C C} = \frac{1}{2\pi\left(100\ \Omega\right)\left(4.0\times10^{-6}\text{ F}\right)} = 4.0\times10^2\text{ Hz} \qquad\qquad \Diamond$$

17.   Determine the maximum magnetic flux through an inductor connected to a standard outlet $(\Delta V_{rms} = 120 \text{ V}, f = 60.0 \text{ Hz})$.

Solution

The AC voltage across a pure inductance and the AC current through it is given by the Ohm's law-like expression $\Delta V_L = I X_L$, where the inductive reactance is $X_L = 2\pi f L = \omega L$. One may use either the maximum voltage and current or the rms values of voltage and current since both $\Delta V_{L, max} = I_{max} X_L$ and $\Delta V_{L, rms} = I_{rms} X_L$ are valid expressions.

As discussed in Section 20.6 in the textbook, the self-inductance of a coil may be expressed in terms of the total magnetic flux through the coil $\Phi_{B, total}$ as

$$L = \frac{\Phi_{B, total}}{I} = \frac{N\Phi_B}{I}$$

where N is the number of turns on the coil, and Φ_B is the flux through a single turn. Thus, the maximum total flux through the inductor is given by

$$\left(\Phi_{B, total}\right)_{max} = LI_{max} = \left(\frac{X_L}{2\pi f}\right)\left(\frac{\Delta V_{L, max}}{X_L}\right) = \frac{\sqrt{2}\left(\Delta V_{L, rms}\right)}{2\pi f}$$

where we have used the fact that $\Delta V_{max} = \sqrt{2}\left(\Delta V_{rms}\right)$ for a sinusoidal voltage.

Hence, if $\Delta V_{L, rms} = 120 \text{ V}$ and $f = 60.0 \text{ Hz}$, we find the maximum flux to be

$$\left(\Phi_{B, total}\right)_{max} = \frac{\sqrt{2}(120 \text{ V})}{2\pi(60.0 \text{ Hz})} = 0.450 \text{ T}\cdot\text{m}^2$$ ◊

23. A 60.0-Ω resistor, a 3.00-μF capacitor, and a 0.400-H inductor are connected in series to a 90.0-V (rms), 60.0-Hz source. Find (a) the voltage drop across the *LC* combination and (b) the voltage drop across the *RC* combination.

Solution

The inductive and capacitive reactances are

$$X_L = 2\pi fL = 2\pi (60.0 \text{ Hz})(0.400 \text{ H}) = 151 \ \Omega$$

and

$$X_C = \frac{1}{2\pi fC} = \frac{1}{2\pi (60.0 \text{ Hz})(3.00 \times 10^{-6} \text{ F})} = 884 \ \Omega$$

The total impedance of the *RLC* series circuit is then

$$Z_{RLC} = \sqrt{R^2 + (X_L - X_C)^2} = \sqrt{(60.0 \ \Omega)^2 + (151 \ \Omega - 884 \ \Omega)^2} = 736 \ \Omega$$

and the current in the series circuit is

$$I_{\text{rms}} = \frac{\Delta V_{RLC,\text{ rms}}}{Z_{RLC}} = \frac{90.0 \text{ V}}{736 \ \Omega}$$

(a) The impedance of the *LC* combination is

$$Z_{LC} = \sqrt{0^2 + (X_L - X_C)^2} = |X_L - X_C| = |151 \ \Omega - 884 \ \Omega| = 733 \ \Omega$$

so the voltage across this section of the series circuit is

$$\Delta V_{LC,\text{ rms}} = I_{\text{rms}} Z_{LC} = \left(\frac{90.0 \text{ V}}{736 \ \Omega}\right)(733 \ \Omega) = 89.6 \text{ V} \qquad \Diamond$$

(b) The impedance of the *RC* section of the series circuit is

$$Z_{RC} = \sqrt{R^2 + (0 - X_C)^2} = \sqrt{(60.0 \ \Omega)^2 + (-884 \ \Omega)^2} = 886 \ \Omega$$

and the voltage across this section of the circuit is

$$\Delta V_{RC,\text{ rms}} = I_{\text{rms}} Z_{RC} = \left(\frac{90.0 \text{ V}}{736 \ \Omega}\right)(886 \ \Omega) = 108 \text{ V} \qquad \Diamond$$

27.   An AC source with a maximum voltage of 150 V and f = 50.0 Hz is connected between points a and d in Figure P21.27. Calculate the rms voltages between points (a) a and b, (b) b and c, (c) c and d, and (d) b and d.

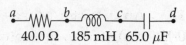

Figure P21.27

Solution

The reactances between points a and d are

$$X_L = 2\pi fL = 2\pi(50.0 \text{ Hz})(0.185 \text{ H}) = 58.1 \text{ }\Omega$$

and $$X_C = \frac{1}{2\pi fC} = \frac{1}{2\pi(50.0 \text{ Hz})(65.0 \times 10^{-6} \text{ F})} = 49.0 \text{ }\Omega$$

and the total impedance between points a and d is

$$Z_{ad} = \sqrt{R_{ad}^2 + \left(X_{L,ad} - X_{C,ad}\right)^2} = \sqrt{(40.0 \text{ }\Omega)^2 + (58.1 \text{ }\Omega - 49.0 \text{ }\Omega)^2} = 41.0 \text{ }\Omega$$

When $\left(\Delta V_{ad}\right)_{rms} = \left(\Delta V_{ad}\right)_{max}/\sqrt{2} = (150 \text{ V})/\sqrt{2}$, the rms current is

$$I_{rms} = \frac{\left(\Delta V_{ad}\right)_{rms}}{Z_{ad}} = \frac{150 \text{ V}}{\sqrt{2}(41.0 \text{ }\Omega)} = 2.58 \text{ A}$$

(a)   Between point a and b,

$$Z_{ab} = \sqrt{R_{ab}^2 + \left(X_{L,ab} - X_{C,ab}\right)^2} = \sqrt{(40.0 \text{ }\Omega)^2 + (0 - 0)^2} = 40.0 \text{ }\Omega$$

and $\left(\Delta V_{ab}\right)_{rms} = I_{rms}Z_{ab} = (2.58 \text{ A})(40.0 \text{ }\Omega) = 103 \text{ V}$ ◊

(b)   For points b and c,

$$Z_{bc} = \sqrt{R_{bc}^2 + \left(X_{L,bc} - X_{C,bc}\right)^2} = \sqrt{(0)^2 + (58.1 \text{ }\Omega - 0)^2} = 58.1 \text{ }\Omega$$

and $\left(\Delta V_{bc}\right)_{rms} = I_{rms}Z_{bc} = (2.58 \text{ A})(58.1 \text{ }\Omega) = 150 \text{ V}$ ◊

(c)   For points c and d,

$$Z_{cd} = \sqrt{R_{cd}^2 + \left(X_{L,cd} - X_{C,cd}\right)^2} = \sqrt{(0)^2 + (0 - 49.0 \text{ }\Omega)^2} = 49.0 \text{ }\Omega$$

and $\left(\Delta V_{cd}\right)_{rms} = I_{rms}Z_{cd} = (2.58 \text{ A})(49.0 \text{ }\Omega) = 127 \text{ V}$ ◊

(d)   Between points b and d,

$$Z_{bd} = \sqrt{R_{bd}^2 + \left(X_{L,bd} - X_{C,bd}\right)^2} = \sqrt{(0)^2 + (58.1 \text{ }\Omega - 49.0 \text{ }\Omega)^2} = 9.15 \text{ }\Omega$$

and $\left(\Delta V_{bd}\right)_{rms} = I_{rms}Z_{bd} = (2.58 \text{ A})(9.15 \text{ }\Omega) = 23.6 \text{ V}$ ◊

31. An inductor and a resistor are connected in series. When connected to a 60-Hz, 90-V (rms) source, the voltage drop across the resistor is found to be 50 V (rms), and the power delivered to the circuit is 14 W. Find (a) the value of the resistance and (b) the value of the inductance.

Solution

(a) The average power delivered to an AC circuit is

$$\mathcal{P}_{av} = I_{rms}(\Delta V_{rms})\cos\phi$$

From the phasor diagram at the right, it is clear that the rms voltage across the resistor is

$$\Delta V_{R, rms} = (\Delta V_{rms})\cos\phi$$

Thus, the power can be written as

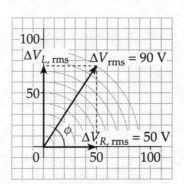

Phasor diagram for an *RL* circuit

$$\mathcal{P}_{av} = I_{rms}(\Delta V_{R, rms})$$

and the rms current in the circuit is $\quad I_{rms} = \dfrac{\mathcal{P}_{av}}{\Delta V_{R, rms}} = \dfrac{14 \text{ W}}{50 \text{ V}} = 0.28 \text{ A}$

Thus, the resistance of the resistor in the circuit is given by

$$R = \frac{\Delta V_{R, rms}}{I_{rms}} = \frac{50 \text{ V}}{0.28 \text{ A}} = 1.8 \times 10^2 \ \Omega \qquad \lozenge$$

(b) Observe from the phasor diagram that $\Delta V_{rms} = \sqrt{\left(\Delta V_{R, rms}\right)^2 + \left(\Delta V_{L, rms}\right)^2}$

The voltage across the inductor is then $\Delta V_{L, rms} = \sqrt{\left(\Delta V_{rms}\right)^2 - \left(\Delta V_{R, rms}\right)^2}$

or $\qquad \Delta V_{L, rms} = \sqrt{(90 \text{ V})^2 - (50 \text{ V})^2} = 75 \text{ V}$

The inductive reactance, $\qquad X_L = \Delta V_{L, rms}/I_{rms}$

is then $\qquad X_L = 75 \text{ V}/0.28 \text{ A} = 2.7 \times 10^2 \ \Omega$

And the self-inductance is $\qquad L = \dfrac{X_L}{\omega} = \dfrac{2.7 \times 10^2 \ \Omega}{2\pi (60 \text{ Hz})} = 0.71 \text{ H} \qquad \lozenge$

37.   A 10.0-Ω resistor, a 10.0-mH inductor, and a 100-μF capacitor are connected in series to a 50.0-V (rms) source having variable frequency. Find the energy delivered to the circuit during one period if the operating frequency is twice the resonance frequency.

Solution

The resonance frequency for a series *RLC* circuit is

$$\omega_0 = 2\pi f_0 = 1/\sqrt{LC}$$

Thus, the operating frequency of interest is

$$\omega = 2\omega_0 = 2/\sqrt{LC}$$

At this frequency, the inductive reactance of the specified circuit is

$$X_L = \omega L = \left(\frac{2}{\sqrt{LC}}\right)L = 2\sqrt{\frac{L}{C}} = 2\sqrt{\frac{10.0\times 10^{-3}\ \text{H}}{100\times 10^{-6}\ \text{F}}} = 20.0\ \Omega$$

and the capacitive reactance is

$$X_C = \frac{1}{\omega C} = \left(\frac{\sqrt{LC}}{2C}\right) = \frac{1}{2}\sqrt{\frac{L}{C}} = \frac{1}{2}\sqrt{\frac{10.0\times 10^{-3}\ \text{H}}{100\times 10^{-6}\ \text{F}}} = 5.00\ \Omega$$

The rms current in the circuit is then

$$I_{\text{rms}} = \frac{\Delta V_{\text{rms}}}{Z} = \frac{\Delta V_{\text{rms}}}{\sqrt{R^2 + (X_L - X_C)^2}} = \frac{50.0\ \text{V}}{\sqrt{(10.0\ \Omega)^2 + (20.0\ \Omega - 5.00\ \Omega)^2}} = 2.77\ \text{A}$$

Therefore, the power supplied to the circuit is

$$\mathcal{P}_{\text{av}} = I_{\text{rms}}^2 R = (2.77\ \text{A})^2 (10.0\ \Omega) = 76.9\ \text{W}$$

and the energy converted in one period $\left(\Delta t = T = 2\pi/\omega = \pi\sqrt{LC}\right)$ is

$$E = \mathcal{P}_{\text{av}}(\Delta t) = (76.9\ \text{W})\pi\sqrt{(10.0\times 10^{-3}\ \text{H})(100\times 10^{-6}\ \text{F})} = 0.242\ \text{J} = 242\ \text{mJ} \qquad \Diamond$$

41. A transformer on a pole near a factory steps the voltage down from 3 600 V (rms) to 120 V (rms). The transformer is to deliver 1 000 kW to the factory at 90% efficiency. Find (a) the power delivered to the primary, (b) the current in the primary, and (c) the current in the secondary.

Solution

(a) The efficiency of the transformer is:

$$\text{eff} = \frac{\left(\mathscr{P}_{av}\right)_{output}}{\left(\mathscr{P}_{av}\right)_{input}} = 0.90$$

Since the factory requires a delivered power (that is, power output from the transformer) of 1 000 kW, the required power input to the primary of the transformer is

$$\left(\mathscr{P}_{av}\right)_{input} = \frac{\left(\mathscr{P}_{av}\right)_{output}}{0.90} = \frac{1\,000\ \text{kW}}{0.90} = 1.1 \times 10^3\ \text{kW} \qquad \Diamond$$

(b) In terms of the primary voltage and current, the input power for the transformer is

$$\left(\mathscr{P}_{av}\right)_{input} = \left(\Delta V_{p,\,rms}\right) I_{p,\,rms}$$

Therefore, the required primary current is

$$I_{p,\,rms} = \frac{\left(\mathscr{P}_{av}\right)_{input}}{\Delta V_{p,\,rms}} = \frac{1.1 \times 10^3\ \text{kW}}{\Delta V_p} = \frac{1.1 \times 10^6\ \text{W}}{3\,600\ \text{V}} = 3.1 \times 10^2\ \text{A} \qquad \Diamond$$

(c) The power output from a transformer is the product of the secondary voltage and current, or

$$\left(\mathscr{P}_{av}\right)_{output} = \left(\Delta V_{s,\,rms}\right) I_{s,\,rms}$$

Hence, the secondary current for this transformer is

$$I_{s,\,rms} = \frac{\left(\mathscr{P}_{av}\right)_{output}}{\Delta V_{s,\,rms}} = \frac{1\,000\ \text{kW}}{\Delta V_{s,\,rms}} = \frac{1.0 \times 10^6\ \text{W}}{120\ \text{V}} = 8.3 \times 10^3\ \text{A} \qquad \Diamond$$

49.   The Sun delivers an average power of $1\,340$ W/m^2 to the top of Earth's atmosphere. Find the magnitudes of $\vec{E}_{max}$ and $\vec{B}_{max}$ for the electromagnetic waves at the top of the atmosphere.

Solution

The average intensity (power per unit area) of an electromagnetic wave may be expressed as

$$\text{Intensity} = \frac{E_{max}B_{max}}{2\mu_0}$$

where E_{max} and B_{max} are the amplitudes of the oscillating electric and magnetic fields associated with the wave.

The ratio of the electric to the magnetic fields in an electromagnetic wave equals the speed of light in a vacuum, or $E_{max}/B_{max} = c$. Thus, the intensity may be written as

$$\text{Intensity} = \frac{\left(cB_{max}\right)B_{max}}{2\mu_0} = \frac{cB_{max}^2}{2\mu_0}$$

The amplitude of the magnetic field associated with the solar radiation at the top of the atmosphere is

$$B_{max} = \sqrt{\frac{2\mu_0\left(\text{Intensity}\right)}{c}} = \sqrt{\frac{2\left(4\pi\times10^{-7}\ \text{T}\cdot\text{m/A}\right)\left(1\,340\ \text{W/m}^2\right)}{3.00\times10^8\ \text{m/s}}} = 3.35\times10^{-6}\ \text{T} \qquad \lozenge$$

The amplitude of the electric field associated with this radiation is then

$$E_{max} = cB_{max} = \left(3.00\times10^8\ \text{m/s}\right)\left(3.35\times10^{-6}\ \text{T}\right) = 1.01\times10^3\ \text{V/m} = 1.01\ \text{kV/m} \qquad \lozenge$$

53.   Infrared spectra are used by chemists to help identify an unknown substance. Atoms in a molecule that are bound together by a particular bond vibrate at a predictable frequency, and light at that frequency is absorbed strongly by the atom. In the case of the C=O double bond, for example, the oxygen atom is bound to the carbon by a bond that has an effective spring constant of $2\,800$ N/m. If we assume that the carbon atom remains stationary (it is attached to other atoms in the molecule), determine the resonant frequency of this bond and the wavelength of light that matches that frequency. Verify that this wavelength lies in the infrared region of the spectrum. (The mass of an oxygen atom is 2.66×10^{-26} kg .)

Solution

Resonance will occur in a oscillating system when the system is excited by a periodic driving force having a frequency equal to the natural frequency of oscillation of the system. The natural frequency of oscillation of an object of mass m on the end of a spring having force constant k is given by

$$f_0 = \frac{1}{2\pi} \sqrt{\frac{k}{m}}$$

Thus, the resonant frequency for the C=O double bond will be

$$f_0 = \frac{1}{2\pi} \sqrt{\frac{k_{effective}}{m_{oxygen \atop atom}}} = \frac{1}{2\pi} \sqrt{\frac{2\,800 \text{ N/m}}{2.66 \times 10^{-26} \text{ kg}}} = 5.2 \times 10^{13} \text{ Hz} \qquad \Diamond$$

The wavelength of light whose electric and magnetic fields oscillate at this frequency is

$$\lambda_0 = \frac{c}{f_0} = \frac{3.00 \times 10^8 \text{ m/s}}{5.2 \times 10^{13} \text{ Hz}} = 5.8 \times 10^{-6} \text{ m} = 5.8 \ \mu\text{m} \qquad \Diamond$$

The infrared region of the electromagnetic spectrum includes wavelengths ranging from $\lambda_{max} \approx 1$ mm down to $\lambda_{min} = 700$ nm $= 0.7 \ \mu$m . Thus, the radiation needed to produce resonance of the C=O double bond lies in the infrared portion of the electromagnetic spectrum. $\qquad \Diamond$

59.   A 200-Ω resistor is connected in series with a 5.0-μF capacitor and a 60-Hz, 120-V rms line. If electrical energy costs \$0.080/kWh, how much does it cost to leave this circuit connected for 24 h?

Solution

The capacitive reactance of the capacitor is

$$X_C = \frac{1}{2\pi fC} = \frac{1}{2\pi(60 \text{ Hz})(5.0 \times 10^{-6} \text{ F})} = 5.3 \times 10^2 \ \Omega$$

Thus, the rms current in the circuit will be

$$I_{rms} = \frac{\Delta V_{rms}}{Z} = \frac{\Delta V_{rms}}{\sqrt{R^2 + (X_L - X_C)^2}} = \frac{120 \text{ V}}{\sqrt{(200 \ \Omega)^2 + (0 - 530 \ \Omega)^2}} = 0.21 \text{ A}$$

The only element in the circuit that produces a net dissipation of energy is the resistor. Hence, the power dissipated in the circuit is

$$\mathcal{P}_{av} = I_{rms}^2 R = (0.21 \text{ A})^2 (200 \ \Omega) = 9.0 \text{ W}$$

and, at a rate of \$0.080/kWh, the cost of operating this circuit for 24 h is

$$cost = E \times rate = \mathcal{P}_{av}(\Delta t) \times rate = (9.0 \text{ W})\left(\frac{1 \text{ kW}}{10^3 \text{ W}}\right)(24 \text{ h})\left(0.080 \ \frac{\$}{\text{kWh}}\right)$$

or $cost = \$0.017 = 1.7$ cents   ◊

65.   One possible means of achieving space flight is to place a perfectly reflecting aluminized sheet into Earth's orbit and to use the light from the Sun to push this solar sail. Suppose such a sail, of area 6.00×10^4 m^2 and mass $6\,000$ kg, is placed in orbit facing the Sun. (a) What force is exerted on the sail? (b) What is the sail's acceleration? (c) How long does it take for this sail to reach the Moon, 3.84×10^8 m away? Ignore all gravitational effects, and assume a solar intensity of $1\,340$ W/m^2. [*Hint*: The radiation pressure by a reflected wave is given by 2(average power per unit area/c).]

Solution

When electromagnetic radiation reflects (at normal incidence) from a perfectly reflecting surface, the momentum imparted to that surface in time Δt is $\Delta p = 2U/c$. Here, U is the energy delivered to the surface during Δt, and c is the speed of light.

From the impulse-momentum theorem, the average force exerted on the surface during this time is

$$F_{av} = \frac{\text{Impulse}}{\Delta t} = \frac{\Delta p}{\Delta t} = \frac{2U}{c(\Delta t)}$$

The intensity, I, of the incident radiation is the power transported per unit area, so the energy delivered to a surface of area A in time Δt is

$$U = \mathcal{P}_{av}(\Delta t) = (I \cdot A)(\Delta t)$$

The average force on the surface is then $F_{av} = \dfrac{2(I \cdot A)(\Delta t)}{c(\Delta t)} = \dfrac{2IA}{c}$

(a)   If $I = 1\,340$ W/m^2, the average force on a perfectly reflecting sail of area $A = 6.00 \times 10^4$ m^2 is

$$F_{av} = \frac{2(1\,340 \text{ W/m}^2)(6.00 \times 10^4 \text{ m}^2)}{3.00 \times 10^8 \text{ m/s}} = 0.536 \text{ N} \qquad \Diamond$$

(b)   From Newton's second law, the average acceleration is

$$a_{av} = \frac{F_{av}}{m} = \frac{0.536 \text{ N}}{6\,000 \text{ kg}} = 8.93 \times 10^{-5} \text{ m/s}^2 \qquad \Diamond$$

(c)   From the uniformly accelerated motion equation $\Delta x = v_0 t + \frac{1}{2}at^2$, with $v_0 = 0$, the time required to reach the moon is

$$t = \sqrt{\frac{2(\Delta x)}{a}} = \sqrt{\frac{2(3.84 \times 10^8 \text{ m})}{8.93 \times 10^{-5} \text{ m/s}^2}} = 2.93 \times 10^6 \text{ s} \left(\frac{1 \text{ day}}{86\,400 \text{ s}} \right) = 33.9 \text{ days} \qquad \Diamond$$

Chapter 22
Reflection and Refraction of Light

NOTES FROM SELECTED CHAPTER SECTIONS

22.1 The Nature of Light

Until the beginning of the 19th century, light was considered to be a stream of particles, emitted by a light source. The first clear demonstration of the wave nature of light was provided in 1801 by Thomas Young, who showed that under appropriate conditions, light exhibits interference behavior. An important development concerning the theory of light was Maxwell's prediction that light is a form of high-frequency electromagnetic waves. On the other hand, Einstein explained the photoelectric effect on the assumption that light is composed of "corpuscles" or discontinuous quanta of energy called photons.

In view of these developments:

> Light must be regarded as having a dual nature; in some cases light acts as a wave, and in others it acts as a particle.

> Light travels in a straight line in a homogenous medium until it strikes a boundary between two different materials.

> Light beams can be represented by a technique called the ray approximation in which a ray of light is shown as a line drawn along the direction of travel of the beam.

22.2 Reflection and Refraction
22.3 The Law of Refraction

Specular reflection occurs when light is reflected from a smooth surface; **diffuse reflection** occurs at rough surfaces. In this and following chapters reflection will be considered to be specular reflection.

A line drawn perpendicular to a surface at the point where an incident ray strikes the surface is called the normal line. **Angles of incidence (θ_1), reflection (θ_1') and refraction (θ_2) are measured relative to the normal**.

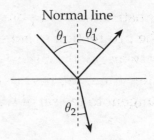

When an incident ray undergoes partial reflection and partial refraction, the incident ray, the reflected ray, the refracted ray and the normal are all in the same plane. For a given angle of incidence, the angle of refraction depends on the optical properties of the media above and below the boundary.

As illustrated in the figure, the angle of reflection (θ_1') is always equal to the angle of incidence (θ_1).

The figure above illustrates a situation in which the speed of light is greater in the medium above the boundary than in the medium below the boundary. In such cases the angle of refraction, as shown, will be smaller than the angle of incidence. This effect is described by Snell's law and expressed in Eqs. 22.3 and 22.8.

The path of a light ray through a refracting surface is reversible.

The index of refraction of a particular material is defined as the ratio of the speed of light in vacuum to its speed in the material in question.

The frequency does not change as light travels from one medium into another.

22.4 Dispersion and Prisms

An important property of the index of refraction (n) is that its value in a refractive material depends on the wavelength of the light. This phenomenon is called dispersion. Since n is a function of the wavelength, when a light beam is incident on the surface of a refracting material, different wavelengths are bent at different angles. *The index of refraction of a given material decreases with increasing wavelength.* This means that blue light bends more than red light, when passing into a refracting material. In a prism, dispersion occurs at one surface as light enters the prism and again at a second surface as it leaves.

22.6 Huygens's Principle

A wave front is a surface which passes through those points in a wave that have both the same phase and amplitude. Huygens's technique for determining the successive positions of a wave front is based on the following geometric construction: All points on a wave front can be considered as point sources for the production of spherical secondary waves (wavelets). The wavelets propagate forward with speeds characteristic of waves in the particular medium. At any later time, the new position of the wave front is found by constructing a surface tangent to the set of wavelets.

22.7 Total Internal Reflection

When light is incident on a boundary between two media, some of the light is always reflected; the remaining light is refracted into the second medium. As the angle of incidence increases, the intensity of the reflected beam increases and that of the refracted beam decreases. If $n_1 > n_2$, as illustrated in the figure, there exists an angle of incidence called the critical angle, θ_c (See Equation 22.9) beyond which no refraction occurs. When $\theta_1 \geq \theta_c$ the incident light is totally reflected back into the medium of greater index of refraction and the intensity of the refracted beam is zero. This phenomenon is called total internal reflection and is possible only when the light is initially traveling in the medium of greater index of refraction ($n_1 > n_2$). When $n_2 > n_1$, some of the incident light is refracted into the second medium for all angles of incidence.

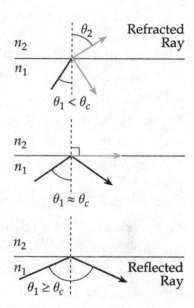

Diagram of total internal reflection
$n_1 > n_2$

EQUATIONS AND CONCEPTS

The **energy of a photon** is proportional to the frequency of the associated electromagnetic wave.

$$E = hf \tag{22.1}$$

$$h = 6.63 \times 10^{-34} \text{ J} \cdot \text{s}$$
(Planck's constant)

The **law of reflection** states that the angle of reflection (the angle measured between the reflected ray and the normal) equals the angle of incidence (the angle between the incident ray and the normal). *The incident ray, reflected ray, and normal line are in the same plane.*

$$\theta_1' = \theta_1 \tag{22.2}$$

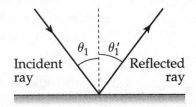

The **index of refraction** of a transparent medium equals the ratio of the speed of light in vacuum to the speed of light in the medium. The index of refraction of a given medium can be expressed as the ratio of the wavelength of light in vacuum to the wavelength in that medium. *The frequency of a wave is characteristic of the source; as light travels from one medium into another of different index of refraction, the frequency remains constant, but the wavelength changes.*

$$n \equiv \frac{\text{speed of light in vacuum}}{\text{speed of light in medium}} = \frac{c}{v} \tag{22.4}$$

$$n = \frac{\lambda_0}{\lambda_n} \tag{22.7}$$

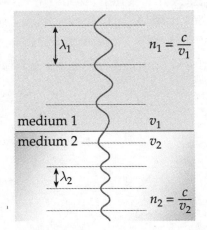

Snell's law of refraction can be expressed in terms of the speeds of light in the media on either side of the refracting surface (Equation 22.3), or in terms of the indices of refraction of the two media (Equation 22.8). *As illustrated in the figure, the angles θ_1 and θ_2 are measured from the normal line to the respective ray.*

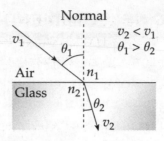

$$\frac{\sin \theta_2}{\sin \theta_1} = \frac{v_2}{v_1} = \text{constant} \qquad (22.3)$$

$$n_1 \sin \theta_1 = n_2 \sin \theta_2 \qquad (22.8)$$

The **critical angle** is the minimum angle of incidence for which total internal reflection can occur. *Total internal reflection is possible only when a light ray is directed from a medium of high index of refraction into a medium of lower index of refraction.*

$$\sin \theta_c = \frac{n_2}{n_1} \text{ for } n_1 > n_2 \qquad (22.9)$$

REVIEW CHECKLIST

▷ When refraction occurs at a plane surface, use Snell's law to calculate the index of refraction of either medium, the angle of refraction or the angle of incidence when the other variables in the equation are known.

▷ Describe the process of dispersion of a beam of white light as it passes through a prism.

▷ Describe the conditions under which total internal reflection of a light ray is possible, and calculate the critical angle for internal reflection at a boundary between two optical media of known indices of refraction. Describe the application of internal reflection to fiber optics techniques.

SOLUTIONS TO SELECTED END-OF-CHAPTER PROBLEMS

5.   Figure P22.5 shows an apparatus used to measure the distribution of the speeds of gas molecules. The device consists of two slotted rotating disks separated by a distance d, with the slots displaced by the angle θ. Suppose the speed of light is measured by sending a light beam toward the left disk of this apparatus. (a) Show that a light beam will be seen in the detector (that is, will make it through both slots) only if its speed is given by $c = \omega d/\theta$,

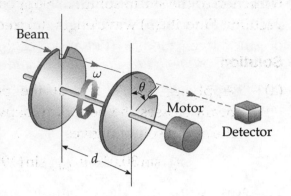

Figure P22.5

where ω is the angular speed of the disks, and θ is measured in radians. (b) What is the measured speed of light if the distance between the two slotted rotating disks is 2.500 m, the slot in the second disk is displaced 1/60 of 1 degree from the slot in the first disk, and the disks are rotating at 5 555 rev/s?

Solution

(a)   If light that passed through the slot on the first disk is to also pass through the slot on the second disk, the second disk must rotate through angle θ in the time interval the light requires to travel distance d between the disks. The duration of this interval is

$$\Delta t = \frac{d}{c}$$

so we must require that $\theta = \omega(\Delta t) = \omega\left(\dfrac{d}{c}\right)$ giving $c = \dfrac{\omega d}{\theta}$ ◊

(b)   If $d = 2.500$ m, $\theta = \left(\dfrac{1}{60}\ \text{deg}\right)\left(\dfrac{2\pi\ \text{rad}}{360\ \text{deg}}\right) = \left(9.259\times10^{-5}\right)\pi$ rad, and the smallest

angular velocity of the rotating disks that allows light to reach the detector is

$\omega = \left(5\,555\ \dfrac{\text{rev}}{\text{s}}\right)\left(\dfrac{2\pi\ \text{rad}}{1\ \text{rev}}\right) = \left(1.111\times10^{4}\right)\pi$ rad/s, then the measured value for

the speed of light is

$$c = \frac{\omega d}{\theta} = \frac{\left[\left(1.111\times10^{4}\right)\pi\ \text{rad/s}\right](2.500\ \text{m})}{\left(9.259\times10^{-5}\right)\pi\ \text{rad}} = 3.000\times10^{8}\ \text{m/s}$$ ◊

9.   A laser beam is incident at an angle of 30.0° to the vertical onto a solution of corn syrup in water. If the beam is refracted to 19.24° to the vertical, (a) what is the index of refraction of the syrup solution? Suppose the light is red, with wavelength 632.8 nm in a vacuum. Find its (b) wavelength, (c) frequency, and (d) speed in the solution.

Solution

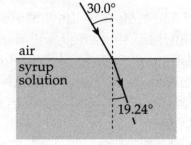

(a)   Applying Snell's law at the point where the beam crosses the boundary between the air and the syrup solution gives

$$n_{air} \sin 30.0° = n_{solution} \sin (19.24°)$$

or $n_{solution} = \dfrac{(1.00)\sin 30.0°}{\sin (19.24°)} = 1.52$ ◊

(b)   In a medium having refractive index n, light whose wavelength in a vacuum is λ_0, will have a wavelength of $\lambda_n = \lambda_0/n$. Thus,

$$\lambda_{solution} = \frac{\lambda_0}{n_{solution}} = \frac{632.8 \text{ nm}}{1.52} = 417 \text{ nm}$$ ◊

(c)   From $v = \lambda f$, we find

$$f = \frac{v_{solution}}{\lambda_{solution}} = \frac{c/n_{solution}}{\lambda_0/n_{solution}} = \frac{c}{\lambda_0} = \frac{3.00 \times 10^8 \text{ m/s}}{632.8 \times 10^{-9} \text{ m}} = 4.74 \times 10^{14} \text{ Hz}$$ ◊

(d)   The speed of light in the syrup solution is

$$v_{solution} = \frac{c}{n_{solution}} = \frac{3.00 \times 10^8 \text{ m/s}}{1.52} = 1.98 \times 10^8 \text{ m/s}$$ ◊

17.   How many times will the incident beam shown in Figure P22.17 be reflected by each of the parallel mirrors?

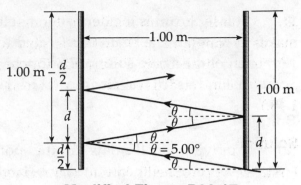

Modified Figure P22.17

Solution

Each time the light beam reflects from a mirror, the angle of incidence and the angle of reflection are equal. The distance between successive points of reflection on either the right-side mirror or the left-side mirror is

$$d = 2\big[(1.00\ \text{m})\tan\theta\big] = 2\big[(1.00\ \text{m})\tan 5.00°\big] = 0.175\ \text{m}$$

Notice that the incident beam just misses the lower end of the right-side mirror, leaving the full length of this mirror available for reflections. Each reflection from this mirror "uses up" a length d of the mirror, and the total number of reflections which may be completed in the full 1.00-m length is

$$N_{right} = \frac{1.00\ \text{m}}{d} = \frac{1.00\ \text{m}}{0.175\ \text{m}} = 5.72 \qquad (\text{or } 5 \text{ full reflections}) \qquad \Diamond$$

The first reflection from the left-side mirror occurs a distance $d/2$ above the lower edge, leaving a mirror length of $(1.00\ \text{m} - d/2)$ to accommodate additional reflections. The number of *additional* reflections which may occur is

$$n_{additional} = \frac{1.00\ \text{m} - d/2}{d} = \frac{0.913\ \text{m}}{0.175\ \text{m}} = 5.22 \qquad (\text{or } 5 \text{ full reflections})$$

Thus, the total number of reflections from this mirror will be

$$N_{left} = 1 + n_{additional} = 1 + 5 = 6 \text{ reflections from the left side mirror} \qquad \Diamond$$

21. The light beam shown in Figure P22.21 makes an angle of 20.0° with the normal line *NN'* in the linseed oil. Determine the angles θ and θ'. (The refractive index for linseed oil is 1.48.)

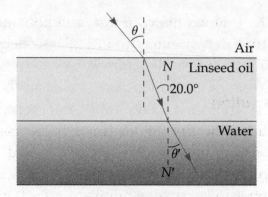

Figure P22.21

Solution

First, we apply Snell's law to the refraction at the air-linseed oil boundary. To do this, realize that the upper and lower surfaces of the linseed oil layer are parallel to each other.

Thus, the two dashed normal lines shown in Figure P22.21 are parallel to each other, and the angle of refraction at the air-linseed oil boundary is 20.0°. Snell's law then gives $n_{air} \sin\theta = n_{oil} \sin 20.0°$, or

$$\theta = \sin^{-1}\left[\frac{n_{oil} \sin 20.0°}{n_{air}}\right] = \sin^{-1}\left[\frac{(1.48)\sin 20.0°}{1.00}\right] = 30.4° \qquad \Diamond$$

Application of Snell's law at the linseed oil-water boundary gives

$$n_{water} \sin\theta' = n_{oil} \sin 20.0°$$

or $\qquad \theta' = \sin^{-1}\left[\frac{n_{oil} \sin 20.0°}{n_{water}}\right] = \sin^{-1}\left[\frac{(1.48)\sin 20.0°}{1.333}\right] = 22.3° \qquad \Diamond$

25. A beam of light both reflects and refracts at the surface between air and glass, as shown in Figure P22.25. If the index of refraction of the glass is n_g, find the angle of incidence, θ_1, in the air that would result in the reflected ray and the refracted ray being perpendicular to each other. [*Hint*: Remember the identity $\sin(90° - \theta) = \cos\theta$.]

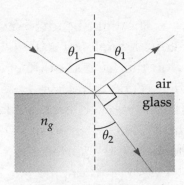

Modified Figure 22.25

Solution

Recall that the angle of reflection is always equal to the angle of incidence. This means that the reflected ray in Figure P22.25 makes angle θ_1 with the dashed normal line shown. Thus, observe in the above figure that $\theta_1 + 90° + \theta_2 = 180°$, so the angle of refraction at the air-glass boundary is given by

$$\theta_2 = 180° - (\theta_1 + 90°) = 90° - \theta_1$$

Application of Snell's law to the refraction at this boundary then gives

$$n_{air} \sin\theta_1 = n_g \sin(90° - \theta_1)$$

Since $\sin(90° - \theta) = \cos\theta$, this necessary condition to have the reflected and refracted rays perpendicular to each other as shown in Figure P22.25 reduces to

$$n_{air} \sin\theta_1 = n_g \cos\theta_1 \qquad \text{or} \qquad \tan\theta_1 = \frac{n_g}{n_{air}}$$

Recognizing that $n_{air} = 1.00$, the required angle of incidence is found to be

$$\theta_1 = \tan^{-1}(n_g) \qquad\qquad\qquad ◊$$

31. A ray of light strikes the midpoint of one face of an equiangular (60°-60°-60°) glass prism ($n = 1.5$) at an angle of incidence of 30°. (a) Trace the path of the light ray through the glass, and find the angles of incidence and refraction at each surface. (b) If a small fraction of light is also reflected at each surface, find the angles of reflection at the surfaces.

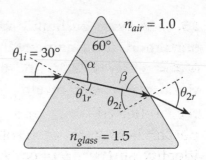

Solution

(a)   The angle of incidence at the first surface is given to be $\theta_{1i} = 30°$ ◊

Snell's law then gives the angle of refraction at the first surface as:

$$\theta_{1r} = \sin^{-1}\left(\frac{n_{air}\sin\theta_{1i}}{n_{glass}}\right) = \sin^{-1}\left(\frac{1.0\sin 30°}{1.5}\right) = 19.47°$$

or rounding to 2 significant figures: $\theta_{1r} = 19°$ ◊

From the sketch, observe that $\theta_{1r} + \alpha = 90°$, or $\alpha = 90° - \theta_{1r}$

Also, $\alpha + \beta + 60° = 180°$ (sum of interior angles in a triangle). Therefore, the angle the refracted ray makes with the second surface is

$$\beta = 120° - \alpha = 120° - (90° - \theta_{1r}) = 30° + \theta_{1r} = 49.47°$$

The angle of incidence at the second surface is then

$$\theta_{2i} = 90° - \beta = 40.53°$$

or rounding to 2 significant figures: $\theta_{2i} = 41°$ ◊

Snell's law gives the angle of refraction at this surface as

$$\theta_{2r} = \sin^{-1}\left(\frac{n_{glass}\sin\theta_{2i}}{n_{air}}\right) = \sin^{-1}\left(\frac{1.5\sin 40.53°}{1.0}\right) = 77°$$ ◊

(b)   From the law of reflection, the angle of reflection is equal to the angle of incidence at each of the surfaces. Thus,

$$\theta_{1,\,reflection} = \theta_{1i} = 30° \quad\text{and}\quad \theta_{2,\,reflection} = \theta_{2i} = 41°$$ ◊

36.   A beam of light is incident from air on the surface of a liquid. If the angle of incidence is 30.0° and the angle of refraction is 22.0°, find the critical angle for the liquid when surrounded by air.

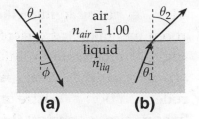

(a)   **(b)**

Solution

When light goes from one material into another having a higher index of refraction, it bends toward the normal line as shown in part (a) of the sketch. It is given that when $\theta = 30.0°$, the angle of refraction in the liquid is $\phi = 22.0°$. Thus, from Snell's law, the index of refraction of the liquid is:

$$n_{liq} = \frac{n_{air} \sin \theta}{\sin \phi} = \frac{(1.00)\sin 30.0°}{\sin 22.0°} = 1.33$$

When light goes from the liquid into the air as shown in part (b), it is going from one material into a second that has a smaller index of refraction. Under these conditions, the refraction is away from the normal $(\theta_2 > \theta_1)$. If the angle of incidence is equal to the critical angle, θ_c, the angle of refraction is $\theta_2 = 90.0°$ (i.e., the refracted ray goes parallel to the surface and never actually enters the second medium).

For any angle of incidence $\theta_1 > \theta_c$, the light is totally internally reflected. Snell's law gives the critical angle as

$$\sin \theta_c = \frac{n_2 \sin 90.0°}{n_1} = \frac{n_2}{n_1}$$

In this case, where the first medium is the liquid, and the second medium is air, the critical angle is found as

$$\sin \theta_c = \frac{n_{air}}{n_{liq}} = \frac{1.00}{1.33} = 0.749 \qquad \text{and} \qquad \theta_c = \sin^{-1}(0.749) = 48.5° \qquad \lozenge$$

41. A room contains air in which the speed of sound is 343 m/s. The walls of the room are made of concrete, in which the speed of sound is 1 850 m/s. (a) Find the critical angle for total internal reflection of sound at the concrete-air boundary. (b) In which medium must the sound be traveling in order to undergo total internal reflection? (c) "A bare concrete wall is a highly efficient mirror for sound." Give evidence for or against this statement.

Solution

Snell's law is valid for any wave disturbance incident on a boundary between two different media. The form of Snell's law most convenient in this case is

$$\frac{\sin \theta_1}{\sin \theta_2} = \frac{v_1}{v_2}$$

Here, θ_1 is the angle of incidence in the first medium (in which the waves propagate at speed v_1), and θ_2 is the angle of refraction in the second medium (where the waves travel at speed v_2).

(a) The critical angle at a boundary between two media is the angle of incidence for which the angle of refraction will be 90.0°. Snell's law then gives

$$\sin \theta_c = \left(\frac{v_1}{v_2}\right)\sin 90.0° = \frac{v_1}{v_2} \quad \text{or} \quad \theta_c = \sin^{-1}\left(\frac{v_1}{v_2}\right)$$

Note that, since no angle has a sine greater than 1.0, the critical angle will exist and total internal reflection will be possible only if $v_1 < v_2$. At the air-concrete boundary, the critical angle for sound waves is

$$\theta_c = \sin^{-1}\left(\frac{v_{air}}{v_{concrete}}\right) = \sin^{-1}\left(\frac{343 \text{ m/s}}{1\,850 \text{ m/s}}\right) = 10.7° \qquad \lozenge$$

(b) For total internal reflection to be possible, the waves must be incident on the boundary from the medium in which they travel the slowest (that is, it is necessary that $v_1 < v_2$). Therefore, when sound waves are incident on a boundary between air and concrete, they must initially be traveling in air if total internal reflection is to be a possibility. $\qquad \lozenge$

(c) Sound waves, initially traveling in air, which strike a concrete-air boundary at any angle of incidence $\theta_1 \geq 10.7°$ will undergo total internal reflection. Thus, the bare concrete wall is an efficient mirror of sound, totally reflecting most of the sound waves striking it. $\qquad \lozenge$

47.   Figure P22.47 shows the path of a beam of light through several layers with different indices of refraction. (a) If $\theta_1 = 30.0°$, what is the angle θ_2 of the emerging beam? (b) What must the incident angle θ_1 be in order to have total internal reflection at the surface between the medium with $n = 1.20$ and the medium with $n = 1.00$?

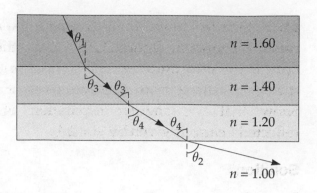

Solution

First, we write Snell's law for the refraction at each of the boundaries.

At the boundary between $n = 1.60$ and $n = 1.40$: $1.60 \sin \theta_1 = 1.40 \sin \theta_3$ **[1]**

At the boundary between $n = 1.40$ and $n = 1.20$: $1.40 \sin \theta_3 = 1.20 \sin \theta_4$ **[2]**

At the boundary between $n = 1.20$ and $n = 1.00$: $1.20 \sin \theta_4 = 1.00 \sin \theta_2$ **[3]**

Since the right side of Equation [1] equals the left side of Equation [2], and the right side of Equation [2] equals the left side of Equation [3], we see that

$$1.60 \sin \theta_1 = 1.00 \sin \theta_2 \qquad \textbf{[4]}$$

Note that Equation [4] is exactly the same as we could have obtained from Snell's law by simply ignoring the presence of the second and third layers of this "sandwich". This will always be true if all the surfaces of the layers in the sandwich of materials are parallel to each other.

(a)   If $\theta_1 = 30.0°$, Equation [4] gives $\theta_2 = \sin^{-1}\left(\dfrac{1.60 \sin 30.0°}{1.00}\right) = 53.1°$ ◊

(b)   At the critical angle of incidence at the $n = 1.20$ and $n = 1.00$ boundary, the refracted ray would be parallel to this boundary (that is, $\theta_2 = 90.0°$). From Equation [4], we find the minimum angle of incidence, $(\theta_1)_{min}$ which would yield total internal reflection at the lower boundary to be

$$(\theta_1)_{min} = \sin^{-1}\left(\frac{1.00 \sin \theta_{2,max}}{1.60}\right) = \sin^{-1}\left(\frac{1.00 \sin 90.0°}{1.60}\right) = 38.7° \qquad ◊$$

51. One technique for measuring the angle of a prism is shown in Figure P22.51. A parallel beam of light is directed onto the apex of the prism so that the beam reflects from opposite faces of the prism. Show that the angular separation of the two reflected beams is given by $B = 2A$.

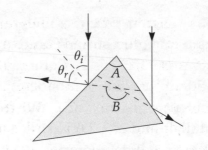

Figure P22.51

Solution

From the sketch at the right, observe that the ray reflecting from the left side of the prism is directed at angle $(\alpha + \beta)$ to the left of the line of the incident light. The ray reflecting from the right side of the prism is at angle $(\gamma + \delta)$ to the right of this direction. Thus, the total angular separation between the two reflected rays is

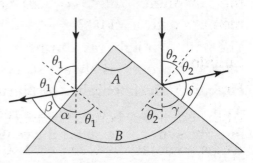

$$B = (\alpha + \beta) + (\gamma + \delta)$$

From the left side of the prism observe that

$$\beta + \theta_1 = 90° \text{ and } \alpha + \beta + 2\theta_1 = 180°$$

Hence, $\quad \alpha = \beta$

Also, looking at the right side of the prism,

$$\delta + \theta_2 = 90° \text{ and } \gamma + \delta + 2\theta_2 = 180° \qquad \text{yielding} \qquad \gamma = \delta$$

We then see that the total angular separation between the reflected rays is

$$B = (\alpha + \beta) + (\gamma + \delta) = (\alpha + \alpha) + (\gamma + \gamma) = 2(\alpha + \gamma) \qquad \text{[1]}$$

Note that the left side of the prism makes angle α with the extension of the line of the incoming light, while the right side of the prism makes angle γ with the extension of the line of the incoming light.. Thus, the angle between these two sides of the prism (that is, the prism angle) is

$$A = \alpha + \gamma \qquad \text{[2]}$$

Substituting Equation [2] into Equation [1] gives the angular separation between reflected rays as

$$B = 2(\alpha + \gamma) = 2A \qquad \Diamond$$

55.   A transparent cylinder of radius $R = 2.00$ m has a mirrored surface on its right half, as shown in Figure P22.55. A light ray traveling in air is incident on the left side of the cylinder. The incident light ray and the exiting light ray are parallel, and $d = 2.00$ m. Determine the index of refraction of the material.

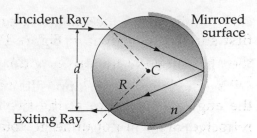

Figure P22.55

Solution

The path of a light ray during a reflection and/or refraction process is always reversible. Thus, if the emerging ray is parallel to the incident ray, the path which the light follows through the cylinder must be symmetric about the center line as shown at the right.

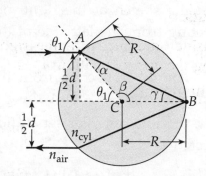

Observe that the angle of incidence of the incoming ray at point A is θ_1. This is also the angle the radius line $\overline{AC}$ makes with the symmetry line at point C.

It is seen that

$$\sin\theta_1 = \frac{d/2}{R} = \frac{d}{2R} = \frac{2.00 \text{ m}}{2(2.00 \text{ m})} = 0.500 \text{ and } \theta_1 = 30.0°$$

Now, observe (at point C) that angle β is the supplement of angle θ_1.

Thus,

$$\beta = 180° - \theta_1 = 180° - 30.0° = 150°$$

From triangle ABC, observe that

$$\alpha + \beta + \gamma = 180°$$

Hence,

$$\alpha + \gamma = 180° - \beta = 30.0°$$

Since triangle ABC in an isosceles triangle (with two sides of length R), the two acute angles (α and γ) in this triangle must be equal.

We then have

$$\alpha + \gamma = \alpha + \alpha = 30.0° \text{ or } \alpha = 15.0°$$

The angle of incidence at point A is $\theta_1 = 30.0°$ while the angle of refraction is $\alpha = 15.0°$. Snell's law then gives the index of refraction of the cylinder material as

$$n_{\text{cylinder}} = \frac{n_{\text{air}} \sin\theta_1}{\sin\alpha} = \frac{(1.00) \sin 30.0°}{\sin 15.0°} = 1.93 \qquad \Diamond$$

59. A light ray incident on a prism is refracted at the first surface, as shown in Figure P22.59. Let ϕ represent the apex angle of the prism and n its index of refraction. Find, in terms of n and ϕ, the smallest allowed value of the angle of incidence at the first surface for which the refracted ray will not undergo total internal reflection at the second surface.

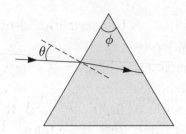

Figure P22.59

Solution

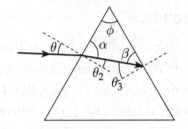

Consider the sketch at the right and apply Snell's law, using the approximation $n_{air} \cong 1$, at the first surface of the prism to find:

$$\sin\theta = \frac{n\sin\theta_2}{n_{air}} = n\sin\theta_2$$

or

$$\theta = \sin^{-1}\left(n\sin\theta_2\right) \qquad\qquad [1]$$

The sum of the interior angles in any triangle is 180°, so we may write

$$\alpha + \beta + \phi = 180° \quad \text{or} \quad \left(90° - \theta_2\right) + \left(90° - \theta_3\right) + \phi = 180°$$

which reduces to $\quad \theta_2 = \phi - \theta_3$

Equation [1] then becomes $\quad \theta = \sin^{-1}\left[n\sin\left(\phi - \theta_3\right)\right]$

Making use of the trigonometric identity $\sin\left(A - B\right) = \sin A\cos B - \cos A\sin B$, this may be rewritten as

$$\theta = \sin^{-1}\left[n\left(\sin\phi\cos\theta_3 - \cos\phi\sin\theta_3\right)\right] \qquad\qquad [2]$$

The angle of incidence at the first surface will have the smallest value for which the light does not undergo total internal reflection at the second surface when θ_3 is equal to the critical angle for the second surface.

That is: $\quad \theta = \theta_{min} \quad$ when $\quad \theta_3 = \theta_c$

and, therefore, $\quad \sin\theta_3 = \sin\theta_c = \frac{n_{air}}{n} = \frac{1}{n}$

Then, $\quad \cos\theta_3 = \sqrt{1 - \sin^2\theta_3} = \sqrt{1 - \frac{1}{n^2}} = \frac{\sqrt{n^2 - 1}}{n}$

and Equation [2] becomes $\quad \theta = \theta_{min} = \sin^{-1}\left[n\left(\frac{\sin\phi\,\sqrt{n^2 - 1}}{n} - \frac{\cos\phi}{n}\right)\right]$

or
$$\theta_{min} = \sin^{-1}\left[\sin\phi\sqrt{n^2-1}-\cos\phi\right]$$ ◊

Note that when $\tan\phi < 1/\sqrt{n^2-1}$, the value of θ_{min} given by this result is negative. This simply means that, when $\tan\phi < 1/\sqrt{n^2-1}$ and the refracted beam is striking the second surface of the prism at the critical angle, the incident beam at the first surface will be on the opposite side of the normal line from what is shown in Figure 22.59.

Chapter 23

Mirrors and Lenses

NOTES FROM SELECTED CHAPTER SECTIONS

In equations and diagrams, the **object distance**, p, is the distance from the object to the mirror (or lens). The **image distance**, q, is the distance from the mirror (or lens) to the location of the image.

23.1 Flat Mirrors

Images formed by a flat mirrors have the following properties:

- The image is as far behind the mirror as the object is in front.

- The image is unmagnified, virtual, and upright.

- The image has *apparent* left-right reversal.

23.2 Images Formed by Spherical Mirrors

A **spherical mirror** is a reflecting surface which has the shape of a segment of a sphere. If the inner surface is reflecting, the mirror is **concave**; if the outer surface of the sphere is reflecting, then the mirror is **convex**.

The **principle axis** of a spherical mirror is along a line drawn from the center (or apex) of the mirror to the **center of curvature** of the mirror. The **focal point** of a concave mirror is on the front side of the mirror at the point along the principle axis at which incoming parallel light rays converge or focus after reflection. For a convex mirror the focal point is on the back side of the mirror at the location from which incoming parallel rays appear to diverge after reflection. The **focal length** (the distance from the mirror to the focal point) is positive for a concave mirror and negative for a convex mirror.

A real image is formed at a point when reflected light actually passes through the point.

A virtual image is formed at a point when light rays appear to diverge from the point.

The object is real when the incident rays diverge from a point located in front of the reflecting surface.

A virtual object exists when the incident rays are converging toward a point located behind the reflecting surface.

23.3 Convex Mirrors and Sign Conventions

A convex mirror is a diverging mirror; light rays from an object (incident on the front side of the mirror) are reflected so as to appear to be coming from the image position (on the back side of the mirror). *Images of real objects formed by convex mirrors are always virtual, upright and smaller than the object. This is true for all object distances.*

A concave mirror is a converging mirror. The nature of an image (real or virtual, upright or inverted, smaller or larger than the object) formed by a concave mirror depends the object distance, p, relative to the focal length, f. Possible object locations and image outcomes are given in the table below:

If the object distance is...	Then the image will be
$p > 2f$	real, inverted, smaller than object
$p = 2f$	real, inverted, same size as object
$f < p < 2f$	real, inverted, larger than object
$p = f$	reflected rays are parallel, no image
$p < f$	virtual, upright, larger than object

A **ray diagram** is a graphical technique (scale drawing) that can be used to determine the location, nature, and relative size of an image formed by a spherical mirror. The position of the object and focal point are shown along the principle axis of the mirror. **See the example ray diagrams for mirrors in Suggestions, Skills and Strategies**.

The point of intersection of any two of the following three rays in a ray diagram for mirrors locates the image:

1.   Ray #1 is drawn from the top of the object parallel to the principle axis and is reflected back through the focal point of a concave mirror. In the case of a convex mirror, the reflected ray appears to diverge from the focal point.

2.   Ray #2 is drawn from the top of the object through the focal point of a concave mirror (toward the focal point of a convex mirror) and is reflected parallel to the principle axis.

3.   Ray #3 is drawn from the top of the object through the center of curvature and is reflected back on itself.

The point at which the rays intersect in the case of a concave mirror (or the point from which they appear to diverge in the case of a convex mirror) locates the top of the image.

23.4   Images Formed by Refraction

A real object is one which is located on the front side of a refracting surface. A virtual object exists when the incident light rays are converging toward a point located on the back side of a refracting surface. The front side is defined as the side of the surface in which light rays originate. The location and character of an image formed by a refracting surface, as determined by using Equations 23.7, 23.8 and 23.9, depend on the shape of the surface (convex, concave, or flat) and the relative values of the indexes of refraction of the media on the two sides of the refracting surface.

The following cases for real objects are illustrated in Suggestions, Skills and Strategies :

(a)   An object located beyond the focal point on the front side of a convex surface where $n_1 < n_2$ will result in a real image on the back side of the refracting surface.

(b)   An object located on the front side of a concave surface where $n_1 < n_2$ will result in a virtual image on the front side of the refracting surface.

(c)   An object located on the front side of a convex surface where $n_1 > n_2$ will result in a virtual image on the front side of the refracting surface.

(d)   An object located on the front side of a flat surface will result in a virtual image on the front side of the surface, with $|q| < p$ if $n_2 < n_1$ and with $|q| > p$ if $n_2 > n_1$.

23.6   Thin Lenses

A converging lens is thicker in the middle than at the rim and has a positive focal length (when surrounded by a medium with an index of refraction smaller than that of the lens). A diverging lens is thicker at the rim and has a negative focal length.

A thin lens has two focal points as illustrated below. Note the reversal of locations of F_1 and F_2 for the two lens types.

The following three rays form the ray diagram for a thin lens:

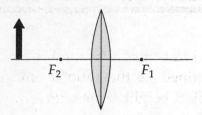

1. Ray #1 is drawn from the object parallel to the principle axis. After being refracted by the lens, this ray passes through focal point F_1 in the case of a converging lens and appears to diverge from F_1 in the case of a diverging lens.

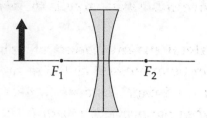

2. Ray #2 is drawn through the center of the lens. This ray continues in a straight line.

3. Ray #3 is drawn through focal point F_2 (toward F_2 in the case of a diverging lens), and emerges from the lens parallel to the principle axis.

In the figure above, consider the arrow to be an object and construct a ray diagram for each of the two cases shown.

See the examples of ray diagrams for thin lenses in Suggestions, Skills and Strategies.

When a combination of two lenses is used to form an image, the image of the first lens is treated as the object for the second lens. The overall magnification of the combination is the product of the magnifications of the separate lenses.

23.7 Lens and Mirror Aberrations

Aberrations are responsible for the formation of imperfect images by lenses and mirrors. Spherical aberration is due to the fact that parallel incident light rays that are different distances from the optical axis are focused at different points following refraction or reflection. Chromatic aberration arises from the fact that light of different wavelengths focuses at different points when refracted by a lens. This occurs because the index of refraction is a function of wavelength.

EQUATIONS AND CONCEPTS

Lateral magnification of a flat mirror is defined as the ratio of image height to the object height. *This ratio always has a value of +1 for a flat mirror since the erect image is always the same size as the object.*

$$M \equiv \frac{\text{image height}}{\text{object height}} = \frac{h'}{h} \qquad (23.1)$$

Lateral magnification of spherical mirrors can be expressed as the negative of the ratio of the image distance to the object distance. *When the image is inverted both M and h' will have negative values.*

$$M = \frac{h'}{h} = -\frac{q}{p} \qquad (23.2)$$

The **focal point of a spherical mirror** is located midway between the center of curvature and the vertex of the mirror. *For a concave mirror, the focal point is in front of the mirror, and f is positive. For a convex mirror, the focal point is back of the mirror, and f is negative.*

$$f = \frac{R}{2} \qquad (23.5)$$

The **mirror equation** is used to determine the location of an image formed by reflection of paraxial rays from a spherical surface.

$$\frac{1}{p} + \frac{1}{q} = \frac{1}{f} \qquad (23.6)$$

A **single spherical refracting surface of radius *R*,** which separates two media whose indices of refraction are n_1 (on the side of the incident rays) and n_2 (on the side of the refracted rays), will form an image of an object. Equations 23.7 and 23.8 are valid regardless of the relative values of n_1 and n_2.

$$\frac{n_1}{p} + \frac{n_2}{q} = \frac{n_2 - n_1}{R} \qquad (23.7)$$

$$M = \frac{h'}{h} = -\frac{n_1 q}{n_2 p} \qquad (23.8)$$

A **flat refracting surface** (when the radius, R, is infinite) will form an image on the same side of the surface as the object regardless of the relative values of n_1 and n_2. *You should review the example ray diagrams and sign conventions for refracting surfaces in* **Suggestions, Skills, and Strategies.**

$$q = -\frac{n_2}{n_1} p \tag{23.9}$$

The **lateral magnification of a thin lens** (Eq. 23.10) has the same form as that of a spherical mirror (Eq. 23.2).

$$M = \frac{h'}{h} = -\frac{q}{p} \tag{23.10}$$

The **thin lens equation** (Eq. 23.11) can be used to find the image location when the focal length is known. Each lens has two focal points as illustrated in the figure below.

$$\frac{1}{p} + \frac{1}{q} = \frac{1}{f} \tag{23.11}$$

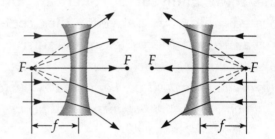

The **lens maker's equation** can be used to calculate the focal length of a thin lens in terms of the radii of curvature of the front and back surfaces and the index of refraction of the lens material. R_1 and R_2 are the respective radii of curvature of the front and back surfaces of the lens. *For a convex lens, R_1 is positive, and R_2 is negative. The reverse is true in the case of a concave lens. If the lens is surrounded by a medium other than air, the index of refraction given in Equation 23.12 must be the ratio of the index of refraction of the lens to that of the surrounding medium.*

$$\frac{1}{f} = (n-1)\left(\frac{1}{R_1} - \frac{1}{R_2}\right) \tag{23.12}$$

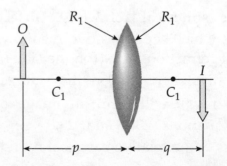

SUGGESTIONS, SKILLS, AND STRATEGIES

A major portion of this chapter is devoted to the development and presentation of equations, which can be used to determine the location and nature of images formed by various optical components acting either singly or in combination. It is essential that these equations be used with the correct algebraic sign associated with each quantity involved. You must understand clearly the sign conventions for mirrors, refracting surfaces, and lenses. The following discussion represents a review of these sign conventions.

SIGN CONVENTIONS FOR SPHERICAL MIRRORS

Equations:
$$\frac{1}{p} + \frac{1}{q} = \frac{1}{f} = \frac{2}{R} \qquad\qquad M = \frac{h'}{h} = -\frac{q}{p}$$

The front side of the mirror is the region on which light rays are incident and reflected.

---

p is + if the object is in front of the mirror (real object).
p is − if the object is in back of the mirror (virtual object).

q is + if the image is in front of the mirror (real image).
q is − if the image is in back of the mirror (virtual image).

Both f and R are + if the center of curvature is in front (concave mirror).
Both f and R are − if the center of curvature is in back (convex mirror).

If M is positive, the image is upright.
If M is negative, the image is inverted.

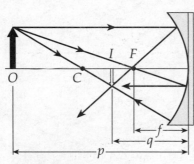

(a) Concave Mirror
$p > 2f$: $q+$, $f+$, $R+$
Image real, inverted, diminished

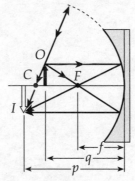

(b) Concave Mirror
$2f > p > f$: $q+$, $f+$, $R+$
Image real, inverted, enlarged

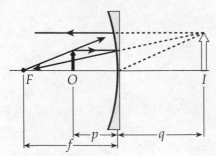

(c) Concave Mirror
$p < f$: $q-$, $f+$, $R+$
Image virtual, upright, enlarged

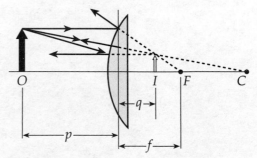

(d) Convex Mirror
$p+$, $q-$, $f-$, $R-$
Image virtual, upright, diminished
for any object distance

SIGN CONVENTIONS FOR REFRACTING SURFACES

Equations:
$$\frac{n_1}{p} + \frac{n_2}{q} = \frac{n_2 - n_1}{R} \qquad M = \frac{h'}{h} = -\frac{n_1 q}{n_2 p}$$

In the following table, the **front** side of the surface is the side **from which the light is incident**.

p	is + if the object is in front of the surface (real object).
p	is − if the object is in back of the surface (virtual object).
q	is + if the image is in back of the surface (real image).
q	is − if the image is in front of the surface (virtual image).
R	is + if the center of curvature is in back of the surface.
R	is − if the center of curvature is in front of the surface.
n_1	refers to the index of refraction of the first medium (before refraction).
n_2	is the index of refraction of the second medium (after refraction).

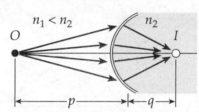

(a) p + (real object)
  q + (real image)
  R + (convex to incident light)

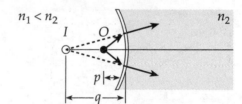

(b) p + (real object)
  q − (virtual image)
  R − (concave to incident light)

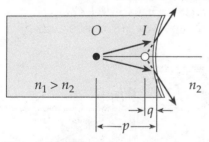

(c) p + (real object)
  q − (virtual image)
  R + (convex to incident light)

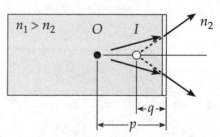

(d) p + (real object)
  q − (virtual image)

SIGN CONVENTIONS FOR THIN LENSES

Equations:
$$\frac{1}{p}+\frac{1}{q}=\frac{1}{f}=(n-1)\left(\frac{1}{R_1}-\frac{1}{R_2}\right) \qquad M=\frac{h'}{h}=-\frac{q}{p}$$

In the following table, the **front** of the lens is the **side from which the light is incident**.

> p is + if the object is in front of the lens.
> p is − if the object is in back of the lens.
>
> q is + if the image is in back of the lens.
> q is − if the image is in front of the lens.
>
> f is + if the lens is thickest at the center.
> f is − if the lens is thickest at the edges.
>
> R_1 and R_2 are + if the center of curvature is in back of the lens.
> R_1 and R_2 are − if the center of curvature is in front of the lens.

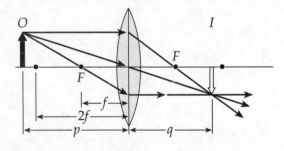

(a) Converging Lens ($f > 0$)
  $p > 2f$: $p+$, $q+$
  Image is real, inverted, diminished

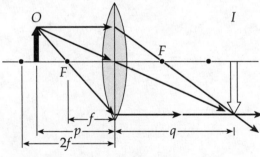

(b) Converging Lens ($f > 0$)
  $2f > p > f$: $p+$, $q+$
  Image is real, inverted, enlarged

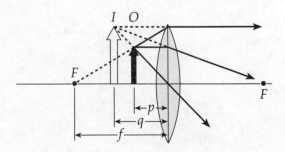

(c) Converging Lens ($f > 0$)
  $p < f$: $p+$, $q-$
  Image is virtual, upright, enlarged

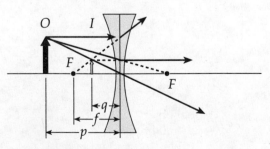

(d) Diverging Lens ($f < 0$)
  $p+$, $q-$
  Image is virtual, upright, diminished

REVIEW CHECKLIST

▷ You should know the sign conventions used with mirrors, lenses and refracting surfaces.

▷ Understand the manner in which the algebraic signs associated with calculated quantities correspond to the nature of the image: real or virtual, upright or inverted.

▷ Calculate the location of the image of a specified object as formed by a flat mirror, spherical mirror, plane refracting surface, spherical refracting surface, or thin lens. Determine the magnification and character of the image in each case.

▷ Construct ray diagrams to determine the location and nature of the image of a given object when the geometric characteristics of the optical device (lens, refracting surface, or mirror) are known.

SOLUTIONS TO SELECTED END-OF-CHAPTER PROBLEMS

3.   A person walks into a room that has, on opposite walls, two plane mirrors producing multiple images. Find the distances from the person to the first three images seen in the left-hand mirror when the person is 5.00 ft from the mirror on the left wall and 10.0 ft from the mirror on the right wall.

Solution

The image formed by a plane mirror is an upright, virtual image located as far behind the mirror as the object is in front of the mirror. The object for the first image formed by the left-hand mirror is the person located 5.00 ft in front of that mirror. This first image (I_{1L}) is located 5.00 ft behind the mirror, or 10.0 ft from the person.   ◊

Images formed by the right hand mirror are located in front of the left-hand mirror and also serve as objects for the left-hand mirror. The person is 10.0 ft in front of the right-hand mirror, so that mirror forms an image (I_{1R}) located 10.0 ft behind it. Since the mirrors are 15.0 ft apart, image I_{1R} is 25.0 ft in front of the left-hand mirror. The left-hand mirror then forms a virtual image (I_{2L}) 25.0 ft behind it. This second image formed by the left-hand mirror is 30.0 ft from the person.   ◊

Image I_{1L} is located $5.0 \text{ ft} + 15.0 \text{ ft} = 20.0 \text{ ft}$ in front of the right-hand mirror and serves as a real object for that mirror. The right-hand mirror then forms an image (I_{2R}) 20.0 ft behind it. This image is $20.0 \text{ ft} + 15.0 \text{ ft} = 35.0 \text{ ft}$ in front of the left-hand mirror. With I_{2R} serving as the object, the left-hand mirror then forms an image (I_{3L}) 35.0 ft behind it and 40.0 ft from the person.   ◊

The process whose beginning is described above continues, with each mirror forming an infinite number of images. Each image formed by a mirror is located farther behind the mirror than the previous image, and it serves as an object for the opposite mirror.

5. At an intersection of hospital hallways, a convex mirror is mounted high on a wall to help people avoid collisions. The mirror has a radius of curvature of 0.550 m. Locate and describe the image of a patient 10.0 m from the mirror. Determine the magnification of the image.

Solution

The radius of curvature of a convex mirror is negative.

Thus, $\qquad\qquad R = -0.550$ m

and the mirror equation gives the image distance as

$$\frac{1}{q} = \frac{2}{R} - \frac{1}{p} = -\frac{2}{0.550 \text{ m}} - \frac{1}{10.0 \text{ m}} = \frac{-2(10.0 \text{ m}) - 0.550 \text{ m}}{(0.550 \text{ m})(10.0 \text{ m})}$$

or $\qquad\qquad q = -0.268$ m

Since $q < 0$, the image is virtual and is located 0.268 m behind the mirror. ◊

The magnification is $\quad M = -\dfrac{q}{p} = -\dfrac{(-0.268 \text{ m})}{10.0 \text{ m}} = +0.026\,8$ ◊

Since, $\qquad\qquad M > 0$, the image is upright

Its size is 0.026 8 times (or 2.68% of) the size of the object. ◊

Use of a convex mirror rather than a flat mirror diminishes the size of the images seen, but it increases the amount of the intersection that can be seen in the mirror. In this case, the "field of view" is $1/0.026\,8 \cong 37$ times greater (in each direction) than it would be if a flat mirror was used.

11.   A 2.00-cm-high object is placed 3.00 cm in front of a concave mirror. If the image is 5.00 cm high and virtual, what is the focal length of the mirror?

Solution

Virtual images formed by concave mirrors are upright images. Hence, the object height, h, and the image height, h', have the same sign, and the magnification is positive.

If $h' = 5.00$ cm when $h = 2.00$ cm, we have $M = \dfrac{h'}{h} = -\dfrac{q}{p}$

or $\dfrac{5.00 \text{ cm}}{2.00 \text{ cm}} = -\dfrac{q}{p}$ and $q = -2.50\, p = -2.50\,(3.00 \text{ cm}) = -7.50 \text{ cm}$

The mirror equation $\dfrac{1}{p} + \dfrac{1}{q} = \dfrac{2}{R} = \dfrac{1}{f}$ then gives

$\dfrac{1}{f} = \dfrac{1}{3.00 \text{ cm}} + \dfrac{1}{-7.50 \text{ cm}}$ or $\dfrac{1}{f} = \dfrac{-7.50 \text{ cm} + 3.00 \text{ cm}}{(3.00 \text{ cm})(-7.50 \text{ cm})}$

The focal length of the mirror is given by

$$f = \dfrac{(3.00 \text{ cm})(-7.50 \text{ cm})}{-7.50 \text{ cm} + 3.00 \text{ cm}} = +5.00 \text{ cm}$$ ◊

15. A man standing 1.52 m in front of a shaving mirror produces an inverted image 18.0 cm in front of it. How close to the mirror should he stand if he wants to form an upright image of his chin that is twice the chin's actual size?

Solution

The image formed when the man stands 1.52 m in front of the mirror $(p = +1.52 \text{ m} = 152 \text{ cm})$ is located on the front side of the mirror. Thus, this is a real image, and the image distance is $q = +18.0 \text{ cm}$. The mirror equation then gives the focal length of the mirror as

$$1/f = 1/p + 1/q$$

or

$$f = \frac{pq}{p+q} = \frac{(152 \text{ cm})(18.0 \text{ cm})}{152 \text{ cm} + 18.0 \text{ cm}} = 16.1 \text{ cm}$$

If the mirror is now to form an upright image, the magnification will be positive. If this upright image is to be twice the size of the object, the magnitude of the magnification will be 2. Thus, we know that $M = +2$

From this, we find

$$M = -\frac{q}{p} = +2 \qquad \text{or} \qquad q = -2p$$

and the mirror equation gives

$$\frac{1}{p} - \frac{1}{2p} = \frac{1}{f} \qquad \text{or} \qquad \frac{1}{2p} = \frac{1}{f}$$

The required object distance is then

$$p = \frac{f}{2} = \frac{16.1 \text{ cm}}{2} = 8.05 \text{ cm} \qquad \Diamond$$

25.   A transparent sphere of unknown composition is observed to form an image of the Sun on its surface opposite the Sun. What is the refractive index of the sphere material?

Solution

The image of the Sun is formed by the refraction of light at the spherical surface facing the Sun. The center of curvature of this surface and the image formed are both located on the back side of this surface. Thus, by the sign convention for refracting surfaces (see Table 23.2 in the textbook), both the radius of curvature and the image distance are positive. If the image is formed on the side of the sphere opposite the Sun, the image distance is

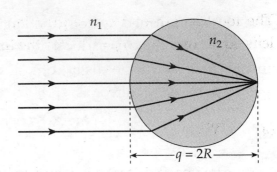

$$q = +(\text{diameter of sphere}) = +2R$$

When light refracts at a spherical surface, going from a medium with refractive index n_1 into a medium having refractive index n_2, the object distance, image distance, and radius of curvature are related by

$$\frac{n_1}{p} + \frac{n_2}{q} = \frac{n_2 - n_1}{R}$$

The distance to the Sun is very large $(p \to \infty)$ in comparison to the size of the sphere, so $n_1/p \to 0$. Also, the sphere is surrounded by air $(n_1 = 1.00)$. Thus, we have

$$0 + \frac{n_2}{q} = \frac{n_2 - 1.00}{R} \qquad \text{or, with } q = 2R, \qquad \frac{n_2}{2R} = \frac{n_2 - 1.00}{R}$$

This gives $n_2 = 2n_2 - 2.00$ and $n_2 = 2.00$ ◊

29. A converging lens has a focal length of 20.0 cm. Locate the images for object distances of (a) 40.0 cm, (b) 20.0 cm, and (c) 10.0 cm. For each case, state whether the image is real or virtual and upright or inverted, and find the magnification.

Solution

The focal length of a converging lens is positive, so $f = +20.0 \text{ cm}$. Solving the thin lens equation, $1/p + 1/q = 1/f$, for the image distance then gives:

$$q = \frac{pf}{p-f} = \frac{p(20.0 \text{ cm})}{p - 20.0 \text{ cm}} \qquad [1]$$

(a) When $\qquad p = +40.0 \text{ cm}$

the image distance is found to be $\qquad q = \dfrac{(40.0 \text{ cm})(20.0 \text{ cm})}{40.0 \text{ cm} - 20.0 \text{ cm}} = +40.0 \text{ cm}$

and the magnification is $\qquad M = -\dfrac{q}{p} = -\dfrac{40.0 \text{ cm}}{40.0 \text{ cm}} = -1.00$

Therefore, the image is real $\qquad (q > 0)$ ◊

inverted $\qquad (M < 0)$ ◊

the same size as the object $\qquad (|M| = 1.00)$ ◊

and located 40.0 cm in back of the lens. ◊

(b) If $\qquad p = f = 20.0 \text{ cm}$

Equation 1 shows that $\qquad q \rightarrow \infty$

In this case, parallel rays leave the lens, and no image is formed. ◊

(c) When $\qquad p = +10.0 \text{ cm}$

Equation 1 gives the image distance $\qquad q = -20.0 \text{ cm}$

The magnification is then $\qquad M = -\dfrac{q}{p} = -\dfrac{(-20.0 \text{ cm})}{10.0 \text{ cm}} = +2.00$

Thus, the image is virtual $\qquad (q < 0)$ ◊

upright $\qquad (M > 0)$ ◊

twice the size of the object $\qquad (|M| = 2.00)$ ◊

and located 20.0 cm in front of the lens. ◊

33.   A transparent photographic slide is placed in front of a converging lens with a focal length of 2.44 cm. The lens forms an image of the slide 12.9 cm from it. How far is the lens from the slide if the image is (a) real? (b) virtual?

Solution

For a converging lens, the focal length is positive. Thus, $f = +2.44$ cm.

(a)   If the image formed by the converging lens is real, it is located on the opposite side of the lens from the real object and the distance between image and object is $p + q = 12.9$ cm as shown in Figure A. The thin lens equation then becomes

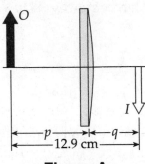

Figure A

$$\frac{1}{p} + \frac{1}{12.9 \text{ cm} - p} = \frac{1}{2.44 \text{ cm}}$$

Finding a common denominator, and simplifying, gives

$$(2.44 \text{ cm})(12.9 \text{ cm} - p + p) = p(12.9 \text{ cm} - p)$$

or $p^2 - (12.9 \text{ cm}) p + 31.5 \text{ cm}^2 = 0$

The quadratic formula then yields two positive solutions, $p = 3.27$ cm and $p = 9.63$ cm. Both are valid object distances for the situation described.   ◊

If $p = 3.27$ cm, then the image distance and the magnification are

$$q = 12.9 \text{ cm} - 3.27 \text{ cm} = 9.63 \text{ cm}$$

and $M = -\dfrac{q}{p} = -\dfrac{9.63 \text{ cm}}{3.27 \text{ cm}} = -2.94$

or the real image is inverted and 2.94 times the size of the object.

If the other solution is used, $p = 9.63$ cm and $q = 3.27$ cm (that is, the values of the image and object distances are interchanged from the above case).

In this case, $M = -\dfrac{q}{p} = -\dfrac{3.27 \text{ cm}}{9.63 \text{ cm}} = -0.340$

so the real image is inverted, and its size is 34.0% that of the object.

(b) If the real object is located inside the focal point, a converging lens forms an upright, magnified, virtual image as shown in Figure B. In this case, the image distance is negative and has a magnitude of $|q| = p + 12.9$ cm.

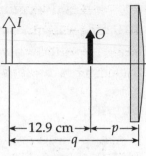

Figure B

Thus, the thin lens equation becomes

$$\frac{1}{p} - \frac{1}{p + 12.9 \text{ cm}} = \frac{1}{2.44 \text{ cm}}$$

which reduces to

$$p^2 + (12.9 \text{ cm})p - 31.5 \text{ cm}^2 = 0$$

This quadratic equation has solutions of $p = +2.10$ cm and $p = -15.0$ cm

Since the slide serves as a real object for the lens, the negative solution must be rejected. Thus, if a virtual image is to be located 12.9 cm from the object, the photographic slide must be positioned 2.10 cm in front of the lens.   ◊

39.   A converging lens is placed 30.0 cm to the right of a diverging lens of focal length 10.0 cm. A beam of parallel light enters the diverging lens from the left, and the beam is again parallel when it emerges from the converging lens. Calculate the focal length of the converging lens.

Solution

When parallel rays enter the divergent lens (L_1), it causes those rays to diverge as though they originated at the focal point on the front side of the lens. This means that it forms a virtual image with image distance $|q_1| = |f_1|$.

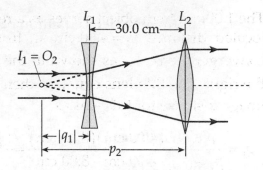

We can also show this by use of the thin lens equation. For parallel incident rays, $p_1 \to \infty$, and $1/p_1 \to 0$. Thus, the thin lens equation becomes

$$0 + \frac{1}{q_1} = \frac{1}{f_1} \qquad \text{giving} \qquad q_1 = f_1 = -10.0 \text{ cm}$$

The virtual image I_1 formed by the divergent lens is located in front of the convergent lens L_2. Therefore, this image serves as a real object for the convergent lens, with an object distance of

$$p_1 = +\left(|q_1| + 30.0 \text{ cm}\right) = +\left(10.0 \text{ cm} + 30.0 \text{ cm}\right) = +40.0 \text{ cm}$$

If the rays emerging from the convergent lens are to be parallel, the object for this lens must be at the focal point located in front of the lens. This means that the focal length of this lens is

$$f_2 = p_2 = +40.0 \text{ cm} \qquad\qquad\qquad \Diamond$$

This can also be seen from the thin lens equation. For parallel emerging rays, $q_2 \to \infty$, and $1/q_2 \to 0$. The thin lens equation then gives

$$\frac{1}{p_2} + 0 = \frac{1}{f_2} \qquad \text{or} \qquad f_2 = p_2 = +40.0 \text{ cm} \qquad\qquad \Diamond$$

43. A 1.00-cm-high object is placed 4.00 cm to the left of a converging lens of focal length 8.00 cm. A diverging lens of focal length −16.0 cm is 6.00 cm to the right of the converging lens. Find the position and height of the final image. Is the image inverted or upright? Real or virtual?

Solution

The 1.00-cm-high object serves as a real object, located distance $p_1 = 4.00$ cm in front of the convergent lens, L_1, as shown in the sketch at the right. The thin lens equation then gives the image distance for this lens as

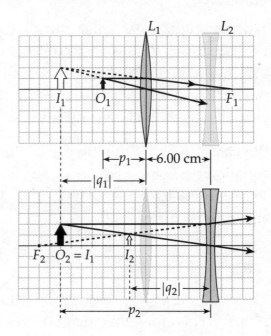

$$q_1 = \frac{p_1 f_1}{p_1 - f_1} = \frac{(4.00 \text{ cm})(8.00 \text{ cm})}{4.00 \text{ cm} - 8.00 \text{ cm}} = -8.00 \text{ cm}$$

The virtual image I_1, formed 8.00 cm in front of the convergent lens, serves as a real object O_2 for the divergent lens L_2. The object distance for this lens is then

$$p_2 = |q_1| + 6.00 \text{ cm} = 14.0 \text{ cm}$$

The image distance for this lens is found from the thin lens equation to be

$$q_2 = \frac{p_2 f_2}{p_2 - f_2} = \frac{(14.0 \text{ cm})(-16.0 \text{ cm})}{14.0 \text{ cm} - (-16.0 \text{ cm})} = -7.47 \text{ cm}$$

The final image formed by this lens combination is a **virtual image** $\left(\text{since } q_2 < 0\right)$ located 7.47 cm in front of the divergent lens, and hence 1.47 cm to the left of the convergent lens. ◊

The overall magnification is
$$M_{\text{overall}} = \frac{h'}{h} = M_1 M_2$$

where L_1 produces magnification
$$M_1 = -\frac{q_1}{p_1} = -\frac{(-8.00 \text{ cm})}{4.00 \text{ cm}} = +2.00$$

and L_2 produces magnification
$$M_2 = -\frac{q_2}{p_2} = -\frac{(-7.47 \text{ cm})}{14.0 \text{ cm}} = +0.534$$

Since the overall magnification is positive, the final image is **upright**. Its height is given by

$$h' = M_{\text{overall}} h = (M_1 M_2) h = \left[(2.00)(0.534) \right](1.00 \text{ m}) = 1.07 \text{ cm} \qquad \Diamond$$

50.   The object in Figure P23.50 is midway between the lens and the mirror. The mirror's radius of curvature is 20.0 cm, and the lens has a focal length of –16.7 cm. Considering only the light that leaves the object and travels first towards the mirror, locate the final image formed by this system. Is the image real or virtual? Is it upright or inverted? What is the overall magnification of the image?

Solution

The ray diagram (*not to scale*) given below shows the formation of the final image. Rays leaving the original object O are caused to converge by the mirror. These rays would form a real image I_m if allowed to do so. However, the lens intercepts and redirects these rays, causing them to diverge as though they came from the upright, virtual, final image I_L.

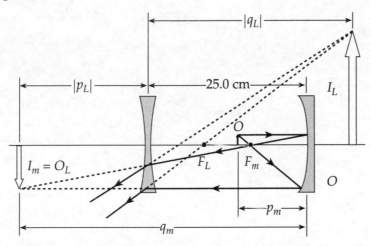

The object distance for the mirror is $p_m = (25.0 \text{ cm})/2 = +12.5 \text{ cm}$

The mirror equation then gives the image distance for the mirror as

$$q_m = \frac{p_m R}{2p_m - R} = \frac{(12.5 \text{ cm})(20.0 \text{ cm})}{2(12.5 \text{ cm}) - 20.0 \text{ cm}} = +50.0 \text{ cm}$$

This real image that the mirror would form serves as the object for the lens. Since it is located behind the lens (that is, on the side opposite the incident light), it is a virtual object, and the object distance for the lens is negative. This object distance is

$$p_L = -(q_m - 25.0 \text{ cm}) = -(50.0 \text{ cm} - 25.0 \text{ cm}) = -25.0 \text{ cm}$$

The image distance for the lens is

$$q_L = \frac{p_L f_L}{p_L - f_L} = \frac{(-25.0 \text{ cm})(-16.7 \text{ cm})}{-25.0 \text{ cm} - (-16.7 \text{ cm})} = -50.3 \text{ cm}$$

Since $q_L < 0$, the final image is a virtual image, located 50.3 cm in front of the lens (that is, on the same side as the incident light) or 25.3 cm behind the mirror.   ◊

The overall magnification produced by this mirror-lens system is given by

$$M_{overall} = M_m M_L$$

where

$$M_m = -\frac{q_m}{p_m} = -\frac{50.0 \text{ cm}}{12.5 \text{ cm}} = -4.00$$

is the magnification by the mirror, and

$$M_L = -\frac{q_L}{p_L} = -\frac{-50.3 \text{ cm}}{-25.0 \text{ cm}} = -2.012$$

is the magnification by the lens. Thus,

$$M_{overall} = (-4.00)(-2.012) = +8.05 \qquad \Diamond$$

Since $M_{overall} > 0$, the final image is upright and 8.05 times the size of the original object. $\Diamond$

55. To work this problem, use the fact that the image formed by the first surface becomes the object for the second surface. Figure P23.55 shows a piece of glass with index of refraction 1.50. The ends are hemispheres with radii 2.00 cm and 4.00 cm, and the centers of the hemispherical ends are separated by a distance of 8.00 cm. A point object is in air, 1.00 cm from the left end of the glass. Locate the image of the object due to refraction at the two spherical surfaces.

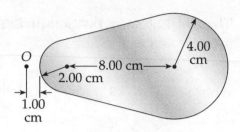

Figure P23.55

Solution

As the light travels left to right through the piece of glass, the center of curvature of the first surface (left end of the glass piece) is located in back of that surface (that is, on the side opposite the incident light). Thus, by the sign convention of Table 23.2 in the text, $R_1 = +2.00 \text{ cm}$. At the second surface (right end of the piece of glass), the center or curvature is in front of the surface or on the same side as the incident light.

Therefore, $$R_2 = -4.00 \text{ cm}.$$

For the refraction at the first surface,

$$p = +1.00 \text{ cm}, \qquad n_1 = n_{air} = 1.00, \qquad \text{and} \qquad n_2 = n_{glass} = 1.50$$

so $\dfrac{n_1}{p} + \dfrac{n_2}{q} = \dfrac{n_2 - n_1}{R}$ becomes $\dfrac{1.00}{1.00 \text{ cm}} + \dfrac{1.50}{q} = \dfrac{1.50 - 1.00}{2.00 \text{ cm}}$

and yields $$q = -2.00 \text{ cm}$$

We see that refraction at this surface forms a virtual image 2.00 cm to the left of the first surface (and hence 16.0 cm to the left of the second surface).

The image formed by the first surface serves as a real object for the second surface (real because it is located in front of that surface). For the refraction at the second surface, we then have

$$p = +16.0 \text{ cm}, \qquad n_1 = n_{glass} = 1.50, \qquad \text{and} \qquad n_2 = n_{air} = 1.00$$

Therefore, $\dfrac{n_1}{p} + \dfrac{n_2}{q} = \dfrac{n_2 - n_1}{R}$ becomes $\dfrac{1.50}{16.0 \text{ cm}} + \dfrac{1.00}{q} = \dfrac{1.00 - 1.50}{-4.00 \text{ cm}}$

This gives $q = +32.0 \text{ cm}$, so the final image is a real image located 32.0 cm in back of (or to the right of) the second surface. ◊

61.   The lens maker's equation for a lens with index n_1 immersed in a medium with index n_2 takes the form

$$\frac{1}{f} = \left(\frac{n_1}{n_2} - 1 \right) \left(\frac{1}{R_1} - \frac{1}{R_2} \right)$$

A thin diverging glass lens (index = 1.50) with $R_1 = -3.00$ m and $R_2 = -6.00$ m is surrounded by air. An arrow is placed 10.0 m to the left of the lens. (a) Determine the position of the image. Repeat part (a) with the arrow and lens immersed in (b) water (index = 1.33); (c) a medium with an index of refraction of 2.00. (d) How can a lens that is diverging in air be changed into a converging lens?

Solution

For the lens described above, the lens maker's equation gives

$$\frac{1}{f} = \left(\frac{1.50}{n_2} - 1 \right) \left(\frac{1}{-3.00 \text{ m}} + \frac{1}{6.00 \text{ m}} \right) = \left(\frac{1.50 - n_2}{n_2} \right) \left(\frac{-1}{6.00 \text{ m}} \right)$$

or $f = \dfrac{n_2 (6.00 \text{ m})}{n_2 - 1.50}$ [1]

(a)   If $n_2 = n_{air} = 1.00$, then $f = \dfrac{1.00 (6.00 \text{ m})}{1.00 - 1.50} = -12.0 \text{ m}$

The thin lens equation gives

$$q = \frac{pf}{p-f} = \frac{(10.0 \text{ m})(-12.0 \text{ m})}{10.0 \text{ m} - (-12.0 \text{ m})} = -5.45 \text{ m}$$

So the lens forms a virtual image 5.45 m to the left of the lens.   ◊

(b)   If $n_2 = n_{water} = 1.33$, then $f = \dfrac{1.33 (6.00 \text{ m})}{1.33 - 1.50} = -46.9 \text{ m}$

The thin lens equation gives

$$q = \frac{pf}{p-f} = \frac{(10.0 \text{ m})(-46.9 \text{ m})}{10.0 \text{ m} - (-46.9 \text{ m})} = -8.24 \text{ m}$$

and the lens forms a virtual image 8.24 m to the left of the lens.   ◊

(c)   If $n_2 = 2.00$, then $\qquad\qquad f = \dfrac{2.00(6.00 \text{ m})}{2.00 - 1.50} = +24.0 \text{ m}$

The thin lens equation gives

$$q = \frac{pf}{p-f} = \frac{(10.0 \text{ m})(24.0 \text{ m})}{10.0 \text{ m} - 24.0 \text{ m}} = -17.1 \text{ m}$$

and the lens forms a virtual image 17.1 m to the left of the lens.   ◊

(d)   From Equation [1], we see that this lens (which is diverging in air) will have a positive focal length and act as a converging lens when $n_2 > 1.50$. Thus, a lens which is diverging in air can be changed into a converging lens by surrounding it with a medium having a refractive index greater than that of the lens material.   ◊

Chapter 24

Wave Optics

NOTES FROM SELECTED CHAPTER SECTIONS

24.1 Conditions for Interference

In order to observe **sustained interference** in light waves, the following conditions must be met:

- The **sources must be coherent**; they must maintain a **constant phase** with respect to each other.

- The **sources must be monochromatic**; they must have identical wavelengths.

- The **superposition principle** must apply.

24.2 Young's Double-Slit Experiment

A schematic diagram (not to scale) illustrating the geometry used in Young's double-slit experiment is shown in the figure below. The two slits S_1 and S_2 serve as coherent monochromatic sources. Light waves from the two sources travel different distances to a point P on the screen. The **path difference is** $\delta = d \sin \theta$.

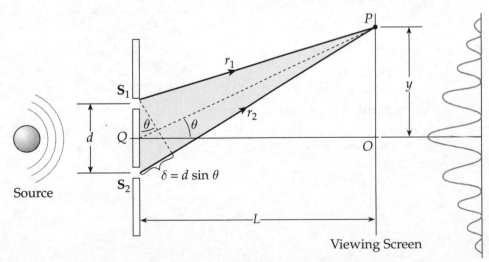

In the figure above, the distribution of relative intensity of light on the screen is shown at right.

Constructive interference (a bright fringe) occurs when the path difference equals either zero or an integer multiple of the wavelength.

Destructive interference (a dark fringe) results when the path difference equals an odd multiple of a half wavelength.

24.3   Change of Phase Due to Reflection

Consider a light wave traveling in a medium with an index of refraction n_1. When partial reflection occurs at the surface of a medium with an index of refraction n_2:

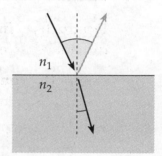

- if $n_1 < n_2$, the reflected ray experiences a phase change of 180°.

- if $n_1 > n_2$, there is no phase change in the reflected ray.

- there is no phase change in the transmitted ray regardless of the relative values of n_1 and n_2.

24.4   Interference in Thin Films

In order to predict constructive or destructive interference in thin films you must consider:

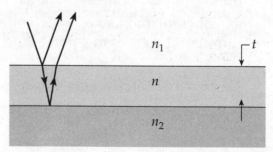

- the difference in path-length traveled by the two interfering waves;

- any expected changes in phase due to reflection;

- the change in wavelength of the light as it enters the film.

There are two general cases to consider (see figure above):

(a) **Reflection resulting in a phase change at only one surface of the film:**

n_1 and n_2 both less than n (phase change at the top surface), or

n_1 and n_2 both greater than n (phase change at the bottom surface)

Constructive interference will occur under these conditions, when the path difference (which equals $2t$) is an odd number of half wavelengths, $2t = (2m+1)\lambda_n/2$. Thus, the film thickness is $t = (m+\frac{1}{2})\lambda_n/2$. *Here, $\lambda_n = \lambda/n$ is the wavelength as measured in the film (the wavelength in vacuum, divided by the index of refraction of the film).* The condition for constructive interference in this case then becomes $t = (m+\frac{1}{2})\lambda/2n$.

Destructive interference will occur under these conditions, when the path difference, $2t$, equals an integer number of wavelengths so that $t = m\lambda/2n$.

(b) **Reflection resulting in phase changes at both top and bottom surfaces of the film (or at neither surface):**

$n_1 < n$ and $n_2 > n$ (phase change at both surfaces), or

$n_1 > n$ and $n_2 < n$ (no phase change at either surface)

In this case, any phase changes that occur are offsetting and interference of the reflected rays depends only on the difference in distance traveled by the two reflected rays and the index of refraction of the film.

Constructive interference will occur when the path difference equals an integer number of wavelengths; the film thickness must be an integer number of half wavelengths, $t = m\lambda/2n$.

Destructive interference in this case will be observed when the path difference equals an odd number of half wavelengths; that is when $t = \left(m+\frac{1}{2}\right)\lambda/2n$.

24.6 Diffraction
24.7 Single-Slit Diffraction

Diffraction occurs when light waves deviate (or spread) from their initial direction of travel when passing through small openings, around obstacles, or by sharp edges.

The diffraction pattern produced by a single narrow slit consists of a broad, intense central band (the central maximum), flanked by a series of narrower and less intense secondary bands (called secondary maxima) alternating with a series of dark bands, or minima.

In the case of single slit diffraction, each portion of the slit acts as a source of waves; and light from one portion of the slit can interfere with light from another portion. The resultant intensity on the screen depends on angle θ, which determines the direction between the perpendicular to the plane of the slit and the direction to a point on the screen.

One type of diffraction, called Fraunhofer diffraction, occurs when the rays reaching the observing screen are approximately parallel.

24.8 The Diffraction Grating

A diffraction grating, consisting of many equally spaced parallel slits, separated by a distance d, will produce an diffraction pattern. There will be a series of principle maxima (bright lines) for each wavelength component in the incident light.

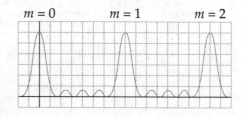

The figure illustrates the case for which the incident light contains a single wavelength component. Although not shown in the figure above, there will also be maxima corresponding to $m = -1, -2, \ldots$ These maxima occur at angles θ, measured from the line perpendicular to the grating, where $m\lambda = d\sin\theta$.

If the incident light contains a second wavelength component, a second series of principle maxima (in general with a different intensity) will be present in the diffraction pattern.

In a given spectral order, denoted by the number m, there will be principle maxima corresponding to each wavelength component incident on the grating.

24.9 Polarization of Light Waves

The electric field vector of a light wave vibrates in a plane perpendicular to the direction of propagation. In the figure the direction of propagation is out of the page. *As illustrated in the top figure, for unpolarized light, the electric field vector vibrates along all directions in the plane perpendicular to the direction of travel with equal probability. However, at a given point and at a particular instant, there is only one resultant electric field direction (shown by the electric field vector $\vec{E}$).*

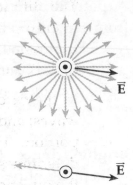

In a linearly polarized wave the electric field vibrates along the same direction at all times at a particular point, as illustrated in the bottom figure.

It is possible to obtain a linearly polarized wave from an unpolarized wave by removing from the unpolarized wave all components except those whose electric field vectors oscillate in a single plane. The plane formed by the direction of vibration of the electric field and the direction of propagation is called the **plane of polarization**.

Three important processes for producing linearly polarized light are:
(1) selective absorption, (2) reflection, and (3) scattering.

Optical activity is the property of certain materials to cause rotation of the plane of polarization as a light beam travels through the material.

EQUATIONS AND CONCEPTS

In **Young's double-slit experiment**, two slits, S_1 and S_2, separated by a distance d, serve as monochromatic coherent sources. The light intensity at any point on the screen is the resultant of light reaching the screen from both slits. *As illustrated in the figure, a point P on the screen can be identified by the angle θ or by the distance y from the center of the screen.*

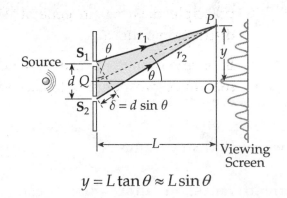

$$y = L \tan\theta \approx L \sin\theta$$

A **path difference** δ arises because waves from S_1 and S_2 travel unequal distances to reach a point on the screen (except the center point, O). *The value of δ determines whether the waves from the two slits arrive in phase or out of phase.*

$$\delta = r_2 - r_1 = d \sin \theta \qquad (24.1)$$

Bright fringes (constructive interference) will appear at points on the screen for which the path difference is equal to integer multiples of the wavelength of the light. The central bright fringe $(\theta_{\text{bright}} = 0, m = 0)$ is called the zeroth-order maximum. The first maxima $(m = \pm1)$ on either side of the zeroth-maxima is called the first-order maxima.

Conditions for double-slit constructive interference:

$$\delta = d \sin \theta_{\text{bright}} = m \lambda \qquad (24.2)$$
$$m = 0, \ \pm 1, \ \pm 2, \ldots$$

$$y_{\text{bright}} = \left(\frac{\lambda L}{d}\right) m \qquad (24.5)$$
$$m = 0, \ \pm 1, \ \pm 2, \ldots$$

m is called the order number

Dark fringes (destructive interference) will appear at points on the screen which correspond to path differences of an odd multiple of half wavelengths. For these points of destructive interference, waves which leave the two slits in phase arrive at the screen 180° (one-half wavelength) out of phase.

Conditions for double-slit destructive interference:

$$\delta = d \sin \theta_{\text{dark}} = \left(m + \tfrac{1}{2}\right) \lambda \qquad (24.3)$$
$$m = 0, \ \pm 1, \ \pm 2, \ldots$$

$$y_{\text{dark}} = \left(\frac{\lambda L}{d}\right)\left(m + \tfrac{1}{2}\right) \qquad (24.6)$$
$$m = 0, \ \pm 1, \ \pm 2, \ldots$$

The **wavelength of light**, λ_n, in a medium having refractive index n, is less than the wavelength in vacuum, λ.

$$\lambda_n = \frac{\lambda}{n} \qquad (24.7)$$

Interference in thin films depends on wavelength, film thickness, and the indices of refraction of the film and surrounding media. *Differences in phase may be due to path difference or phase change upon reflection.*

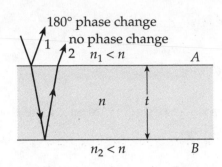

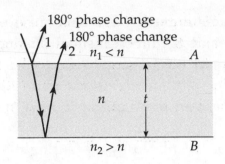

Case (I)	**Phase change at only one film surface** ($n_1 < n$ and $n_2 < n$) or ($n_1 > n$ and $n_2 > n$) Indices of refraction of media on both sides of the film are less than that of the film (figure above left) or both greater than that of the film.	**Constructive interference(Case I)** $$2nt = \left(m + \tfrac{1}{2}\right)\lambda \qquad (24.9)$$ $$(m = 0, 1, 2, \ldots)$$ **Destructive interference(Case I),** $$2nt = m\lambda \qquad (24.10)$$ $$(m = 0, 1, 2, \ldots)$$

Case (II) **Phase changes at both or neither surface** ($n_1 < n < n_2$ or $n_1 > n > n_2$)
Film is between two media either of which has an index of refraction greater than that of the film and the other a smaller index (figure above right).

Constructive interference(Case II)
$$2nt = m\lambda$$
$$(m = 0, 1, 2, \ldots)$$

Destructive interference(Case II)
$$2nt = \left(m + \tfrac{1}{2}\right)\lambda$$
$$(m = 0, 1, 2, \ldots)$$

In **single-slit diffraction**, the total phase difference between waves from the top and bottom portions the slit will depend on the angle θ which determines the direction to a point on the screen. The pattern consists of a broad central bright band and a series of less intense and narrower side bands.

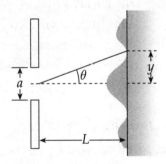

The **general condition for destructive interference** (dark band) at a given point on the screen can be stated in terms of the angle θ_{dark}. Equation 24.11 gives the angle θ_{dark} for which the intensity will be zero (or minimum).

For destructive interference:
$$\sin\theta_{\text{dark}} = m\frac{\lambda}{a} \qquad (24.11)$$
$$m = \pm 1, \ \pm 2, \ \pm 3, \ldots$$

A **diffraction grating** has many narrow and equally-spaced parallel slits. The grating produces a diffraction pattern with a series of maxima (bright lines) for each different wavelength in the incident light. When the grating spacing (d) is known, the wavelength (λ) can be calculated by measuring the angle (θ_{bright}) for a given spectral line

$$d \sin \theta_{bright} = m\lambda \qquad (24.12)$$

$$m = 0, 1, 2, 3, \ldots$$

m is the order number of the diffraction pattern.

Malus's law states that the fraction of initially polarized light (from a polarizer) that will be transmitted by a second sheet of polarizing material (the analyzer) depends on the square of the cosine of the angle θ between the transmission axis of the polarizer and that of the analyzer. From this expression, note that the transmitted intensity is a maximum when the transmission axes are parallel ($\theta = 0°$ or $180°$). When the transmission axes are perpendicular to each other, the light is completely absorbed by the analyzer, and the transmitted intensity is zero.

$$I = I_0 \cos^2 \theta \qquad (24.13)$$

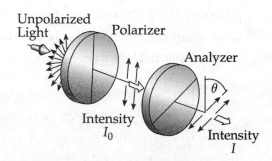

I_0 is the intensity of the beam incident on the analyzer after being transmitted by the polarizer.

The **polarizing angle** is the angle of incidence for which light incident from air and reflected from a surface of index of refraction, n, will be completely polarized with the direction of polarization parallel to the surface. Under this condition, the transmitted light will be partially polarized. Equation 24.14 is known as **Brewster's law.**

$$n = \tan \theta_p \qquad (24.14)$$

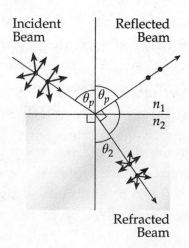

SUGGESTIONS, SKILLS, AND STRATEGIES

THIN FILM INTERFERENCE PROBLEMS

1. Identify the thin film from which interference effects are being observed.

2. The type of interference (constructive or destructive) that occurs in a specific situation is determined by the phase relationship between that portion of the wave reflected at the upper surface of the film and that portion reflected at the lower surface of the film.

3. The phase difference between the two portions of the wave is determined by the physical path difference and also by any phase changes which may occur at either surface upon reflection. **You must consider both potential contributions to the total phase difference.**

    Path difference — The wave reflected from the lower surface of the film has to travel a distance equal to **twice the thickness of the film** before it returns to the upper surface of the film where it interferes with that portion of the wave reflected at the upper surface.

    Phase change upon reflection — In addition to path difference, reflections may change the predicted interference results. When a wave traveling in a particular medium reflects off a surface having a higher index of refraction than the one in which it was initially traveling, a 180° phase shift occurs. This has the same effect as if the wave lost $\frac{1}{2}\lambda$. Phase changes due to reflection must be considered in addition to the extra distance traveled by the reflected wave.

4. **When path difference and phase changes upon reflection are both taken into account**, the interference will be constructive if the two waves are out of phase ("out of step") by an integer multiple of λ_n (the wavelength as measured in the film). Destructive interference will occur when they differ in phase by an odd number of half-wavelengths.

REVIEW CHECKLIST

▷ Describe Young's double-slit experiment to demonstrate the wave nature of light. Account for the phase difference between light waves from the two sources as they arrive at a given point on the screen. State the conditions for constructive and destructive interference in terms of each of the following: path difference (δ), distance from center of screen (y), and the angle between the perpendicular bisector of the two sources and the location on the screen (θ).

▷ Account for the conditions of constructive and destructive interference in thin films considering both path difference and any expected phase changes due to reflection.

▷ Describe Fraunhofer diffraction produced by a single slit. Determine the positions of the minima in a single-slit diffraction pattern.

▷ Describe qualitatively the polarization of light by selective absorption, reflection, and scattering. Also, make appropriate calculations using Brewster's law and Malus's law.

▷ Determine the positions of the principle maxima (or calculate wavelength values) in spectra formed by a diffraction grating.

SOLUTIONS TO SELECTED END-OF-CHAPTER PROBLEMS

5. In a location where the speed of sound is 354 m/s, a 2 000-Hz sound wave impinges on two slits 30.0 cm apart. (a) At what angle is the first maximum located? (b) If the sound wave is replaced by 3.00-cm microwaves, what slit separation gives the same angle for the first maximum? (c) If the slit separation is 1.00 μm, what frequency of light gives the same first maximum angle?

Solution

When waves pass through a pair of parallel slits, the slits act as a set of coherent sources. The waves from these sources interfere constructively (that is, produce maximum net disturbances) when the distance from the observer to one slit is an integral number of wavelengths greater than the distance to the other slit. From the sketch at the right, observe that the difference in path lengths is $\delta = d\sin\theta$, where d is the spacing between slits and θ is measured from the line perpendicular to the plane of the slits. Thus, maxima will occur in those directions where $\delta = d\sin\theta = m\lambda$,

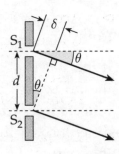

or $\qquad \theta = \sin^{-1}(m\lambda/d)\qquad$ with $\qquad m = 0,\ \pm 1,\ \pm 2,\ \pm 3,\ \ldots$

(a) The wavelength of the sound waves is $\lambda = v/f = (354 \text{ m/s})/2\,000 \text{ Hz} = 0.177 \text{ m}$

With $d = 30.0$ cm, the first order $(m = 1)$ maximum will occur at

$$\theta = \sin^{-1}\left[\frac{1(0.177 \text{ m})}{0.300 \text{ m}}\right] = 36.2° \qquad \Diamond$$

(b) If microwaves of wavelength $\lambda = 3.00$ cm are incident on the slits, the slit spacing required to produce the $m = 1$ maximum at $\theta = 36.2°$ is

$$d = \frac{m\lambda}{\sin\theta} = \frac{1(3.00 \text{ cm})}{\sin 36.2°} = 5.08 \text{ cm} \qquad \Diamond$$

(c) When $d = 1.00$ μm, the wavelength of light that will produce the $m = 1$ maximum at $\theta = 36.2°$ is

$$\lambda = \frac{d\sin\theta}{m} = \frac{(1.00 \times 10^{-6} \text{ m})\sin 36.2°}{1} = 5.91 \times 10^{-7} \text{ m}$$

The frequency of this light is $f = \dfrac{c}{\lambda} = \dfrac{3.00 \times 10^8 \text{ m/s}}{5.91 \times 10^{-7} \text{ m}} = 5.08 \times 10^{14} \text{ Hz} \qquad \Diamond$

13. Radio waves from a star, of wavelength 250 m, reach a radio telescope by two separate paths as shown in Figure P24.13. One is a direct path to the receiver, which is situated on the edge of a cliff by the ocean. The second is by reflection off the water. The first minimum of destructive interference occurs when the star is 25.0° above the horizon. Find the height of the cliff. (Assume no phase change on reflection.)

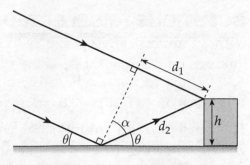

Modified Figure P24.13

Solution

If the star is at angle θ above the horizon, the incident rays will be at angle θ above the horizontal as shown in the sketch. The law of reflection then requires that the reflected ray travel at angle θ above the horizontal. Observe from the sketch that the difference in the distances the direct ray and the reflected ray have traveled when they arrive at the receiver is

$$\delta = d_2 - d_1 = d_2 - d_2 \sin \alpha = d_2 (1 - \sin \alpha)$$

From the sketch, note that $\qquad \theta + 90° + \alpha + \theta = 180° \quad$ or $\quad \alpha = 90° - 2\theta$

Thus, when $\theta = 25.0°$, we have $\alpha = 40.0°$ and $\qquad \delta = d_2 (1 - \sin 40.0°)$

If we assume no phase shift on reflection, this difference in path lengths must equal $\lambda/2 = 125$ m when the first minimum of intensity caused by destructive interference occurs. Hence, we now have

$$125 \text{ m} = d_2 (1 - \sin 40.0°) \qquad \text{or} \qquad d_2 = \frac{125 \text{ m}}{1 - \sin 40.0°} = 350 \text{ m}$$

Then, looking at the sketch again, we see that the height of the cliff is

$$h = d_2 \sin \theta = (350 \text{ m}) \sin 25.0° = 148 \text{ m}$$

◊

19. A possible means for making an airplane invisible to radar is to coat the plane with an antireflective polymer. If radar waves have a wavelength of 3.00 cm, and the index of refraction of the polymer is $n = 1.50$, how thick would you make the coating?

Solution

The concept of an antireflective coating is to produce destructive interference in the waves reflecting from the first and second surfaces of the coating. Assuming the refractive index of the polymer is less than that of the material making up the body of the airplane, waves reflecting from both surfaces of the coating will experience 180° phase shifts. When this is true, the conditions for constructive and destructive interference are reversed from what they are when the waves experience a phase shift at only one of the surfaces. Thus, the condition for destructive interference in this case is

$$2nt = \left(m + \frac{1}{2} \right) \lambda \qquad \text{where} \qquad m = 0, 1, 2, \ldots$$

For the thinnest coating meeting this condition, $m = 0$. Then, the required thickness of the coating is

$$t = \frac{\lambda}{4n} = \frac{3.00 \text{ cm}}{4(1.50)} = 0.500 \text{ cm} \qquad \qquad \Diamond$$

Basing your defensive strategy on a coating such as this would be very risky. The enemy could modify its radar to use a wavelength of 1.50 cm. Then, your polymer coating would make the airplanes very good reflectors by producing constructive interference in the reflected radar waves.

23. An air wedge is formed between two glass plates separated at one edge by a very fine wire, as in Figure P24.22. When the wedge is illuminated from above by 600-nm light, 30 dark fringes are observed. Calculate the radius of the wire.

Figure P24.22

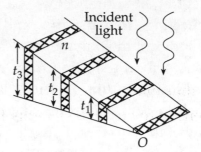

Solution

A thin film of air, of varying thickness, exists between the two glass plates. Light reflecting from the upper surface of this air film (as it attempts to go from glass into air) does not experience a phase reversal. However, light reflecting from the lower surface (as it attempts to go from air into the lower glass plate) does experience a phase reversal. Thus, the two reflected rays have a 180° phase difference (equivalent to a path difference of $\lambda_{air}/2$) due to the difference in the nature of these reflections.

Assuming normal incidence, the total difference in optical path lengths is then

$$\delta_{total} = 2t + \frac{\lambda_{air}}{2} = 2t + \frac{\lambda}{2n_{air}}$$

For destructive interference, the total difference in optical paths must be an odd number of half-wavelengths of the light in the air film. That is:

$$\delta_{total} = (2m+1)\frac{\lambda_{air}}{2} = (2m+1)\frac{\lambda}{2n_{air}} \quad m = 0,1,2,\ldots \quad \textbf{[Destructive Interference]}$$

Therefore, the thicknesses of the air film which yield dark fringes are found from

$$\delta_{total} = 2t + \frac{\lambda}{2n_{air}} = (2m+1)\frac{\lambda}{2n_{air}}$$

or $\qquad t_m = \frac{m\lambda}{2n_{air}} \quad m = 0,1,2,\ldots$

Counting the $m = 0$ order along the edge of contact between the glass plates, the order number of the thirtieth observed dark fringe is $m = 29$. At this point, the thickness of the air film is $t_{29} = 2r$, where r is the radius of the wire. Hence, we have

$$r = \frac{t_{29}}{2} = \frac{29\,\lambda}{4\,n_{air}} = \frac{29\left(600 \times 10^{-9} \text{ m}\right)}{4(1.00)} = 4.35 \times 10^{-6} \text{ m} = 4.35 \ \mu\text{m} \qquad \Diamond$$

27.   A thin film of MgF_2 ($n = 1.38$) with thickness 1.00×10^{-5} cm is used to coat a camera lens. Are any wavelengths in the visible spectrum intensified in the reflected light?

Solution

At each surface of the thin film on the camera lens, the light is going from one medium into another having a higher index of refraction (air to MgF_2 at the first surface and MgF_2 to glass, with $n_{glass} > 1.38$, at the second surface).

Thus, a light ray reflecting from the first surface and a ray reflecting from the second surface both undergo phase reversals upon reflection, leaving zero net phase difference due to reflections.

The only phase difference in the two reflected rays is that due to difference in actual path lengths. Assuming normal incidence, this is $\delta_{total} = 2t$ where t is the thickness of the film. For constructive interference, this total path difference must be an integral number of wavelengths of the light in the film:

$$\delta_{total} = 2t = m\,\lambda_n = \frac{m\,\lambda}{n_{film}} \qquad m = 1, 2, 3, \ldots \qquad \text{[Constructive Interference]}$$

If $t = 1.00 \times 10^{-5}$ cm $= 1.00 \times 10^{-7}$ m $= 100$ nm and $n_{film} = 1.38$,   the   wavelengths intensified (that is, undergoing constructive interference) in the reflected light are

$$\lambda = \frac{2\,n_{film}\,t}{m} = \frac{2(1.38)(100 \text{ nm})}{m} = \frac{276 \text{ nm}}{m} \qquad \text{where} \quad m = 1, 2, 3, \ldots$$

All of these wavelengths are below the range of wavelengths visible to the human eye $(400 \text{ nm} \le \lambda_{visible} \le 700 \text{ nm})$.

Thus, no wavelengths in the visible spectrum are intensified in the reflected light. $\Diamond$

31.   Light of wavelength 587.5 nm illuminates a single slit of width 0.75 mm. (a) At what distance from the slit should a screen be placed if the first minimum in the diffraction pattern is to be 0.85 mm from the central maximum? (b) Calculate the width of the central maximum.

Solution

Light passing through different portions of a single slit interferes in such a way as to produce maxima and minima in intensity on a screen located at distance L beyond the slit as shown in the sketch at the right. The minima or dark fringes occur where

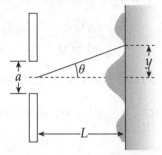

$$\sin\theta = m\lambda/a \quad \text{with} \quad m = \pm 1, \pm 2, \pm 3, \dots$$

When the slit width, a, is very large in comparison to the wavelength, λ, of the light, the angles θ are very small. In this case, $\sin\theta \approx \tan\theta = y/L$ so the locations of the minima on the screen are given by

$$y_m = mL\left(\frac{\lambda}{a}\right) \quad \text{with} \quad m = \pm 1, \pm 2, \pm 3, \dots$$

(a)   If $a = 0.75$ mm and $\lambda = 587.5$ nm, the slit to screen distance required to produce the first $(m = 1)$ minimum at 0.85 mm from the central maximum is

$$L = \frac{y_1}{1}\left(\frac{a}{\lambda}\right) = \left(0.85 \times 10^{-3}\ \text{m}\right)\left(\frac{0.75 \times 10^{-3}\ \text{m}}{587.5 \times 10^{-9}\ \text{m}}\right) = 1.1\ \text{m} \qquad \Diamond$$

(b)   The central maximum extends from the first minimum on one side of the symmetry line to the first minimum on the other side of this line. Thus, its width on the screen is

$$\Delta y = y_1 - y_{-1} = 2y_1 = 2(0.85\ \text{mm}) = 1.7\ \text{mm} \qquad \Diamond$$

37.   The hydrogen spectrum has a red line at 656 nm and a violet line at 434 nm. What angular separation between these two spectral lines is obtained with a diffraction grating that has 4 500 lines/cm ?

Solution

The spacing between adjacent slits on this diffraction grating is

$$d = \frac{1.00 \text{ cm}}{4\,500} = \frac{1.00 \times 10^{-2} \text{ m}}{4\,500} = \frac{1.00 \times 10^{7} \text{ nm}}{4\,500} = 2.22 \times 10^{3} \text{ nm}$$

From the grating equation, $d \sin \theta = m\lambda$

the angular position of the maximum of order m for wavelength λ is

$$\theta = \sin^{-1} (m\lambda / d)$$

Thus, the angular separation between the mth order maxima of the two given wavelengths is

$$\Delta\theta = \sin^{-1} \left(\frac{m\,\lambda_{red}}{d} \right) - \sin^{-1} \left(\frac{m\,\lambda_{violet}}{d} \right) = \sin^{-1} \left[\frac{m(656 \text{ nm})}{2.22 \times 10^{3} \text{ nm}} \right] - \sin^{-1} \left[\frac{m(434 \text{ nm})}{2.22 \times 10^{3} \text{ nm}} \right]$$

or $\Delta\theta = \sin^{-1} \left[m(0.295) \right] - \sin^{-1} \left[m(0.195) \right]$

For the first order $(m = 1)$ this gives $\Delta\theta = 5.91°$ ◊

For the second order $(m = 2)$ this gives $\Delta\theta = 13.2°$ ◊

For the third order $(m = 3)$ this gives $\Delta\theta = 26.5°$ ◊

While using this grating, it is not possible to see complete orders for $m > 3$.

43.   A beam of 541-nm light is incident on a diffraction grating that has 400 lines/mm. (a) Determine the angle of the second-order ray. (b) If the entire apparatus is immersed in water, determine the new second-order angle of diffraction. (c) Show that the two diffracted rays of parts (a) and (b) are related through the law of refraction.

Solution

The spacing between adjacent slits on this diffraction grating is

$$d = \frac{1.00 \text{ mm}}{400} = \frac{1.00 \times 10^{-3} \text{ m}}{400} = \frac{1.00 \times 10^{6} \text{ nm}}{400} = 2.50 \times 10^{3} \text{ nm}$$

If the index of refraction of the medium surrounding the diffraction grating is n, the wavelength of the light in that medium is $\lambda_n = \lambda/n$ where λ is the wavelength in a vacuum. The grating equation, $d \sin \theta = m \lambda_n$, then gives the angle of the second-order $(m = 2)$ diffracted ray as

$$\theta = \sin^{-1}\left(\frac{2 \lambda_n}{d}\right) = \sin^{-1}\left(\frac{2 \lambda}{d n}\right)$$

(a)   If the surrounding medium is air, then $n = n_{air} = 1.00$, and the angle for the second-order ray is

$$\theta_{air} = \sin^{-1}\left(\frac{2 \lambda}{d n_{air}}\right) = \sin^{-1}\left[\frac{2(541 \text{ nm})}{(2.50 \times 10^{3} \text{ nm})(1.00)}\right] = 25.6° \qquad \Diamond$$

(b)   When the surrounding medium is water, then $n = n_{water} = 1.33$, and the angular position of the second-order ray is

$$\theta_{water} = \sin^{-1}\left(\frac{2 \lambda}{d n_{water}}\right) = \sin^{-1}\left[\frac{2(541 \text{ nm})}{(2.50 \times 10^{3} \text{ nm})(1.33)}\right] = 19.0° \qquad \Diamond$$

(c)   From the grating equation, $d \sin \theta = m \lambda_n = m(\lambda/n)$

Thus, we have $n \sin \theta = \frac{m \lambda}{d} = \text{constant}$

when both m and λ are held constant. Therefore, $n_{air} \sin \theta_{air} = n_{water} \sin \theta_{water}$ or the angles of parts (a) and (b) satisfy the law of refraction (that is, Snell's law). $\Diamond$

_____

47. The index of refraction of a glass plate is 1.52. What is the Brewster's angle when the plate is (a) in air? (b) in water? (See Problem 51.)

Solution

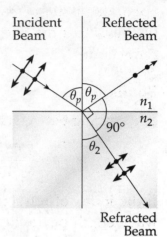

Incident Beam   Reflected Beam

n_1

$90°$ n_2

θ_2

Refracted Beam

If, when light is incident on a surface, the reflected beam and the refracted beam are perpendicular to each other as shown in the figure at the right, the reflected beam will be completely polarized with the electric field vector parallel to the surface. The angle of incidence, θ_p, at which this occurs is called the polarizing angle or the Brewster's angle.

From the figure, observe that $\qquad \theta_p + 90° + \theta_2 = 180°$

or $\qquad\qquad\qquad\qquad\qquad \theta_2 = 90° - \theta_p$

Thus, $\qquad\qquad\qquad\qquad \sin\theta_2 = \sin\left(90° - \theta_p\right) = \cos\theta_p$

Applying Snell's law to the refraction at the surface then gives

$$n_1 \sin\theta_p = n_2 \sin\theta_2 = n_2 \cos\theta_p$$

or $\qquad\qquad\qquad\qquad \tan\theta_p = \dfrac{\sin\theta_p}{\cos\theta_p} = \dfrac{n_2}{n_1}$

(a) If the glass plate is in air, $\qquad n_1 = 1.00$ and $n_2 = 1.52$

The Brewster's angle is then $\quad \theta_p = \tan^{-1}\left(\dfrac{n_2}{n_1}\right) = \tan^{-1}\left(\dfrac{1.52}{1.00}\right) = 56.7°$ ◊

(b) When there is water above the glass plate, $\quad n_1 = 1.33$ and $n_2 = 1.52$

and Brewster's angle is $\qquad \theta_p = \tan^{-1}\left(\dfrac{n_2}{n_1}\right) = \tan^{-1}\left(\dfrac{1.52}{1.33}\right) = 48.8°$ ◊

53.   Three polarizing plates whose planes are parallel are centered on a common axis. The directions of the transmission axes relative to the common vertical direction are shown in Figure P24.53. A linearly polarized beam of light with plane of polarization parallel to the vertical reference direction is incident from the left onto the first disk with intensity $I_i = 10.0$ units (arbitrary). Calculate the transmitted intensity I_f when $\theta_1 = 20.0°$, $\theta_2 = 40.0°$, and $\theta_3 = 60.0°$. [*Hint:* Make repeated use of Malus's law.]

Solution

When linearly polarized light of incident intensity I_0 passes through an analyzer whose transmission axis is rotated at angle θ to the plane of polarization of the incident light, the emerging light is linearly polarized in a direction parallel to the transmission axis of the analyzer. The intensity of the emerging light is given by

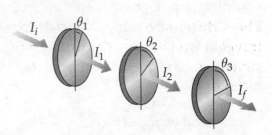

Figure P24.53

$$I = I_0 \cos^2 \theta \qquad \text{[Malus's law]}$$

The first analyzer in Figure P24.53 has light of intensity I_i, polarized in the vertical plane, incident on it. The emerging light has intensity

$$I_1 = I_i \cos^2 \theta_1$$

The transmission axis of the second analyzer in the figure is at angle $\theta = \theta_2 - \theta_1$ from the plane of polarization of the light incident on it. The intensity of the light emerging from this analyzer is

$$I_2 = I_1 \cos^2 (\theta_2 - \theta_1) = I_i \cos^2 \theta_1 \cos^2 (\theta_2 - \theta_1)$$

The third analyzer has its transmission axis at angle $\theta = \theta_3 - \theta_2$ from the plane of polarization of the incoming light, so the emerging light has intensity

$$I_f = I_2 \cos^2 (\theta_3 - \theta_2) = I_i \cos^2 \theta_1 \cos^2 (\theta_2 - \theta_1) \cos^2 (\theta_3 - \theta_2)$$

If $I_i = 10.0$ units, $\theta_1 = 20.0°$, $\theta_2 = 40.0°$, and $\theta_3 = 60.0°$, the final emergent intensity is

$$I_f = (10.0 \text{ units}) \cos^2 (20.0°) \cos^2 (20.0°) \cos^2 (20.0°) = 6.89 \text{ units} \qquad \Diamond$$

63.   Figure P24.63 shows a radio wave transmitter and a receiver, both $h = 50.0$ m above the ground and $d = 600$ m apart. The receiver can receive signals directly from the transmitter, and indirectly, from signals that bounce off the ground. If the ground is level between the transmitter and receiver, and a $\lambda/2$ phase shift occurs upon reflection, determine the longest wavelengths that interfere (a) constructively and (b) destructively.

Solution

The difference in the distances traveled by the direct and the indirect signals from the transmitter to the receiver is

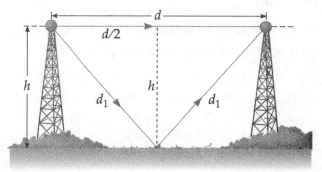

$$\Delta d = 2d_1 - d$$

where $d_1 = \sqrt{(d/2)^2 + h^2} = \sqrt{d^2 + 4h^2}/2$

Transmitter   Receiver

or $\Delta d = \sqrt{d^2 + 4h^2} - d$

Modified Figure P24.63

In addition, the reflected signal experiences a 180° phase shift (equivalent to a path difference of $\lambda/2$) upon reflection. Thus, the total difference in path lengths of the two signals is

$$\delta = \Delta d + \lambda/2 = \sqrt{d^2 + 4h^2} - d + \lambda/2$$

(a)   For constructive interference, we must have $\delta = m\lambda$, where m is any integer. This gives

$$\sqrt{d^2 + 4h^2} - d = \left(m - \frac{1}{2}\right)\lambda \qquad \text{or} \qquad \lambda = \frac{2\left[\sqrt{d^2 + 4h^2} - d\right]}{2m - 1}$$

The longest wavelength satisfying this condition occurs for $m = 1$. With $h = 50.0$ m and $d = 600$ m, we find

$$\lambda_{max} = 2\left[\sqrt{(600 \text{ m})^2 + 4(50.0 \text{ m})^2} - 600 \text{ m}\right] = 16.6 \text{ m} \qquad \Diamond$$

(b)   For destructive interference, it is necessary that $\delta = (m + 1/2)\lambda$, where m is a positive integer. Thus, $\sqrt{d^2 + 4h^2} - d = m\lambda$. For the longest finite wavelength that can produce destructive interference, $m = 1$, which gives

$$\lambda_{max} = \sqrt{d^2 + 4h^2} - d = \sqrt{(600 \text{ m})^2 + 4(50.0 \text{ m})^2} - 600 \text{ m} = 8.28 \text{ m} \qquad \Diamond$$

67. A diffraction pattern is produced on a screen 140 cm from a single slit, using monochromatic light of wavelength 500 nm. The distance from the center of the central maximum to the first-order maximum is 3.00 mm. Calculate the slit width. [*Hint:* Assume that the first-order maximum is halfway between the first- and second-order minima.]

Solution

Dark fringes or minima in a single slit diffraction pattern exist at angles θ for which

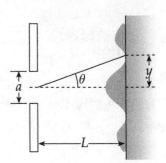

$$\sin \theta = m\lambda/a \quad \text{with} \quad m = \pm 1, \pm 2, \pm 3, \ldots$$

Here, θ is measured from the line passing through the center of the slit and perpendicular to the plane of the slit as shown in the figure, a is the width of the slit, and λ is the wavelength of the light illuminating the slit.

When $a \gg \lambda$, the angles where the minima occur are quite small, and we may use the approximation $\sin \theta \approx \tan \theta = y/L$. The screen position, measured from the location of the central maximum, where the minimum of order m occurs is then seen to be

$$y_m = mL\left(\frac{\lambda}{a}\right)$$

Assuming the first-order maximum is halfway between the first- and second-order minima, its distance from the central maximum is

$$\left(y_{\max}\right)_1 = y_1 + \frac{1}{2}\left(y_2 - y_1\right) = \frac{y_2 + y_1}{2} = \frac{1}{2}\left[2L\left(\frac{\lambda}{a}\right) + \left(\frac{\lambda}{a}\right)\right] = \frac{3}{2}L\left(\frac{\lambda}{a}\right)$$

If it is observed that $\left(y_{\max}\right)_1 = 3.00$ mm when $\lambda = 500$ nm and $L = 140$ cm, the width of the slit must be

$$a = \frac{3L\lambda}{2\left(y_{\max}\right)_1} = \frac{3(1.40 \text{ m})\left(500 \times 10^{-9} \text{ m}\right)}{2\left(3.00 \times 10^{-3} \text{ m}\right)} = 3.50 \times 10^{-4} \text{ m} = 0.350 \text{ mm} \qquad \Diamond$$

Chapter 25

Optical Instruments

NOTES FROM SELECTED CHAPTER SECTIONS

25.2 The Eye

The following terms are used in describing the image-forming mechanism of the eye:

- **Cornea**: transparent structure through which light enters the eye.

- **Aqueous humor**: clear liquid behind the cornea.

- **Pupil**: an opening in the iris, dilates and contracts to regulate light.

- **Iris**: colored portion of the eye, muscular diaphragm that controls pupil size.

- **Retina, rods and cones**: back surface of eye, where images are formed.

- **Accommodation**: changing the focal length of the eye by action of the ciliary muscle.

- **Near point**: the closest distance for which the lens can accommodate to focus a sharp image on the retina (average value of around 25 cm).

- **Far point**: the greatest distance for which the lens in a relaxed eye can focus a clear image on the retina (in a normal eye the far point is at infinity).

- **Hyperopia** (farsightedness): condition caused either when the eyeball is too short or when the ciliary muscle is unable to change the shape of the lens enough to form a properly focused image of a nearby object.

- **Myopia** (nearsightedness): condition caused either when the eye is longer than normal or when the maximum focal length of the lens is insufficient to produce a clearly focused image of a distant object.

- **Presbyopia**: a reduction in accommodation ability resulting in the symptoms of farsightedness.

- **Astigmatism**: light from a point source produces a line image on the retina.

- **Diopters**: a measure of the power of a lens which equals the inverse of the focal length in meters.

25.4   The Compound Microscope

The overall magnification of a compound microscope of length L is equal to the product of the lateral magnification produced by the objective lens of focal length f_0 (< 1 cm) and the angular magnification produced by the eyepiece lens of focal length f_e (a few centimeters). The two lenses, separated by a distance greater than either f_0 or f_e form an inverted, virtual image of an object.

25.5   The Telescope

There are two fundamentally different types of telescopes, both designed to aid in viewing distant objects, such as the planets in our solar system.

- The **refracting telescope** uses a combination of objective and eyepiece lenses to form an enlarged image. The two lenses are separated by a distance $L = f_e + f_o$.

- The **reflecting telescope** uses a curved mirror and an eyepiece lens to form an enlarged image.

For each type telescope, the angular magnification of the telescope equals the ratio of the focal length of the objective to the focal length of the eyepiece.

25.6   Resolution of Single-Slit and Circular Apertures

The wave nature of light limits, by diffraction, the minimum distance between objects which can be distinguished by an optical system.

Rayleigh's criterion determines the limiting condition for resolution of adjacent sources. Under this criterion two sources are just resolved when they are spaced so that the central maximum of the diffraction pattern of one source is located at the position of the first minimum of the diffraction pattern of the second source. Under this condition they are separated by the minimum spacing for which they can be seen as separate sources.

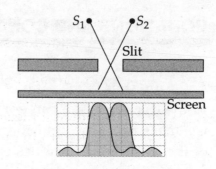

The smallest angular separation that can be resolved is proportional to the wavelength of light used by the instrument and inversely proportional to the slit width or the diameter of a circular aperture.

The resolving power of a diffraction grating increases with the order number in the spectrum and is proportional to the number of lines illuminated in the grating.

25.7   The Michelson Interferometer

The Michelson interferometer can be used to measure distances to within a fraction of a wavelength of light. This is accomplished by splitting a light beam into two parts and then recombining the beams to form an interference pattern.

EQUATIONS AND CONCEPTS

The *f*-number is the ratio of the focal length to the diameter of a camera lens. This value is a measure of the light concentrating power of the lens and determines what is called the speed of the lens. *A fast lens has a small f-number; usually a short focal length and a large diameter.*

$$f - \text{number} \equiv \frac{f}{D} \tag{25.1}$$

The **power of a lens is measured in diopters** and equals the inverse of the focal length measured in meters. The correct algebraic sign must be used with f (+ for converging lenses, and – for diverging lenses). *Optometrists and ophthalmologists usually prescribe lenses measured in diopters.*

$$\mathcal{P} = \frac{1}{f}$$

The **angular magnification of a simple magnifier** is the ratio of the angle subtended by an object when a lens is in use to the angle subtended by the object when it is placed at the near point of the eye (25 cm) with no lens. The magnifier increases the angular size of the object (the size of the angle subtended by the object at the eye).

$$m \equiv \frac{\theta}{\theta_0} \tag{25.2}$$

When the image is at the near point of the eye, the angular magnification of a simple magnifier is maximum. In this case $q = -25\text{ cm}$.

$$m_{\text{max}} = 1 + \frac{25\text{ cm}}{f} \tag{25.5}$$

When the **image is at infinity** (most relaxed for the eye) the angular magnification is minimum. This is the expression for the magnification of a simple lens when the object is placed at the focal point of the lens.

$$m = \frac{25\text{ cm}}{f} \tag{25.6}$$

A **compound microscope** contains an objective lens of short focal length f_o and an eyepiece of focal length f_e. The two lenses are separated by a distance L. When an object is located just beyond the focal point of the objective, the two lenses in combination form an enlarged, virtual, and inverted image of overall magnification, M. The negative sign indicates the inverted nature of the image.

$$M = M_1 m_e = -\frac{L}{f_o}\left(\frac{25 \text{ cm}}{f_e}\right) \qquad (25.7)$$

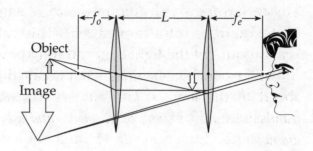

The **angular magnification of a telescope** is equal to the ratio of the focal length of the objective to the focal length of the eyepiece. The two converging lenses are separated by a distance equal to the sum of their focal lengths.

$$m = \frac{\theta}{\theta_0} = \frac{f_o}{f_e} \qquad (25.8)$$

Rayleigh's criterion states the condition for the resolution of the images of two closely spaced sources.

> For a **slit** the angular separation between sources (in radians) must be greater than the ratio of the wavelength of the light to the slit width, a.

$$\theta_{\min} \cong \frac{\lambda}{a} \qquad (25.9)$$
(for a slit)

> For a **circular aperture**, the minimum angular separation which can be resolved depends on the wavelength of the light and the diameter of the aperture, D.

$$\theta_{\min} = 1.22\frac{\lambda}{D} \qquad (25.10)$$
(for a circular aperture)

The **resolving power of a diffraction grating** increases as the number of lines illuminated, N, increases; also, the resolving power is proportional to the order number, m, in which the spectrum is observed.

$$R \equiv \frac{\lambda}{\lambda_2 - \lambda_1} = \frac{\lambda}{\Delta\lambda} \qquad (25.11)$$

$$R = Nm \qquad (25.12)$$

From Equation 25.11, it can be seen that a grating with a high resolving power can distinguish small differences between adjacent wavelengths. Equation 25.12 shows that in the zeroth order $(m = 0)$, the resolution $R = 0$, and all wavelengths are indistinguishable. Now consider viewing a spectrum in the second order $(m=2)$ when 5 000 lines of the grating are illuminated. By combining Equation 25.11 and Equation 25.12 we find that $\lambda / \Delta\lambda = Nm$. So in this example, the grating will produce a spectrum in which wavelengths differing in value by 1 part in 10 000 can be resolved.

Resolving power of a grating

REVIEW CHECKLIST

▷ Define the *f*-number of a camera lens and relate this criterion to shutter speed.

▷ Describe the geometry of the optical components for each of several optical instruments: simple magnifier, compound microscope, reflecting telescope, and refracting telescope. Also, calculate the magnifying power for each instrument.

▷ Determine whether or not two sources under a given set of conditions are resolvable as defined by Rayleigh's criterion.

▷ Describe the technique employed in the Michelson interferometer for precise measurement of length based on known values for the wavelength of light.

▷ Understand what is meant by the resolving power of a grating, and calculate the resolving power of a grating under specified conditions.

SOLUTIONS TO SELECTED END-OF-CHAPTER PROBLEMS

7. A certain type of film requires an exposure time of 0.010 s with an $f/11$ lens setting. Another type of film requires twice the light energy to produce the same level of exposure. What f-stop does the second type of film need with the 0.010-s exposure time?

Solution

The energy delivered to the film by the light striking it is $\Delta E = \mathscr{P}(\Delta t) = IA(\Delta t)$, where I is the intensity of the light, A is the area of the film that is exposed, and Δt is the exposure time.

Since the exposure time and the area of film exposed (that is the size of the image) is the same with the two films, we see that if we are to double the energy delivered to the film, the intensity of the light reaching the film must be doubled.

The intensity of light reaching the film is proportional to the reciprocal of the square of the f-stop setting. (See the statement of Problem 6 in Chapter 25 of the textbook). That is,

$$I = \frac{constant}{(f-\text{number})^2}$$

Thus, if the intensity I_2 of the light exposing the second film is to be double the intensity I_1 used to expose the first film, it is necessary that

$$\frac{constant}{(f-\text{number})_2^2} = 2\left[\frac{constant}{(f-\text{number})_1^2}\right]$$

or $(f-\text{number})_2^2 = \dfrac{(f-\text{number})_1^2}{2} = \dfrac{(11)^2}{2} = 61$, giving $(f-\text{number})_2 = 7.8$

This means you should use the $f/8$ setting on your camera for the new film. ◊

11. The accommodation limits for Nearsighted Nick's eyes are 18.0 cm and 80.0 cm. When he wears his glasses, he is able to see faraway objects clearly. At what minimum distance is he able to see objects clearly?

Solution

In order for Nick to see distant objects clearly, the lens in his glasses must form an erect, virtual image of very distant objects $(p \to \infty)$ at the far point of his eyes (that is, the image distance must be $q = -80.0$ cm).

The thin lens equation then gives the required focal length for the lens in his glasses as

$$\frac{1}{f} = \frac{1}{p} + \frac{1}{q} = \frac{1}{\infty} + \frac{1}{-80.0 \text{ cm}} \text{ or } f = -80.0 \text{ cm}$$

When Nick uses his glasses to observe nearby objects, the lens must form an erect, virtual image of the closest objects viewed clearly at the near point of his eye. That is, when $p = p_{\min}$, it is necessary that $q = -18.0$ cm. The thin lens equation then gives the minimum viewing distance as

$$\frac{1}{p} = \frac{1}{f} - \frac{1}{q} \text{ or } p_{\min} = \frac{fq}{q - f} = \frac{(-80.0 \text{ cm})(-18.0 \text{ cm})}{-18.0 \text{ cm} - (-80.0 \text{ cm})} = 23.2 \text{ cm} \qquad \Diamond$$

16. A person is to be fitted with bifocals. She can see clearly when the object is between 30 cm and 1.5 m from the eye. (a) The upper portions of the bifocals (Fig. P25.16) should be designed to enable her to see distant objects clearly. What power should they have? (b) The lower portions of the bifocals should enable her to see objects comfortably at 25 cm. What power should they have?

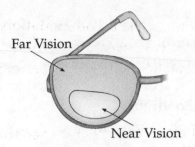

Far Vision

Near Vision

Figure P25.16

Solution

(a) To correct a nearsighted eye (fails to see distant objects clearly), the corrective lens should form an upright, virtual image of the most distant objects $(p \to \infty)$ at the far point of the eye (that is, $q = -1.5$ m in this case). The required focal length for the upper portion of the corrective lens is found from the thin lens equation as

$$\frac{1}{f_u} = \frac{1}{p} + \frac{1}{q} = \frac{1}{\infty} + \frac{1}{-1.5 \text{ m}} \qquad \text{or} \qquad f_u = -1.5 \text{ m}$$

Thus, the power of the upper portion of the bifocals should be

$$\mathcal{P}_u = \frac{1}{f_u} = \frac{1}{-1.5 \text{ m}} = -0.67 \text{ diopters} \qquad \diamond$$

(b) The lower portion of the bifocals must allow the person to see objects located 25 cm from the eye (at the normal near point) clearly. Since the nearest object the unaided eye can focus on is 30 cm from the eye, the corrective lens must form an upright, virtual image 30 cm in front of the eye ($q = -30$ cm) when the object distance is $p = +25$ cm. The thin lens equation then gives the required focal length of the lower portion of the corrective lens as

$$\frac{1}{f_L} = \frac{1}{25 \text{ cm}} + \frac{1}{-30 \text{ cm}} = \frac{+1}{150 \text{ cm}} \qquad \text{or} \qquad f_L = +150 \text{ cm} = +1.5 \text{ m}$$

The required power of the lower portion of the lens is then

$$\mathcal{P}_L = \frac{1}{f_L} = \frac{1}{+1.5 \text{ m}} = +0.67 \text{ diopters} \qquad \diamond$$

21.   A leaf of length h is positioned 71.0 cm in front of a converging lens with a focal length of 39.0 cm. An observer views the image of the leaf from a position 1.26 m behind the lens, as shown in Figure P25.21. (a) What is the magnitude of the lateral magnification (the ratio of the image size to the object size) produced by the lens? (b) What angular magnification is achieved by viewing the image of the leaf rather than viewing the leaf directly?

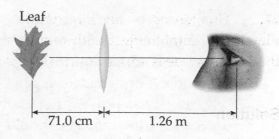

Figure P25.21

Solution

(a)   With a focal length of $f = +39.0$ cm and an object distance of $p = +71.0$ cm, the thin lens equation gives the image distance as

$$q = \frac{pf}{p-f} = \frac{(71.0\ \text{cm})(39.0\ \text{cm})}{71.0\ \text{cm} - 39.0\ \text{cm}} = +86.5\ \text{cm}$$

Thus, the image of the leaf is 86.5 cm to the right of the lens or 39.5 cm in front of the observer's eye. The lateral magnification by the lens is

$$M = \frac{h'}{h} = -\frac{q}{p} = -\frac{86.5\ \text{cm}}{71.0\ \text{cm}} = -1.22 \qquad \Diamond$$

(b)   If the unaided eye viewed the leaf, of height h, directly from a distance of $d = 126\ \text{cm} + 71.0\ \text{cm} = 197\ \text{cm}$, the angular size of the leaf would be

$$\theta_0 = \frac{h}{d} = \frac{h}{197\ \text{cm}}$$

If instead, the eye views the image formed by the lens (of height $h' = |M|h = 1.22h$) from a distance of $d' = 126\ \text{cm} - 86.5\ \text{cm} = 39.5\ \text{cm}$, the angular size of the object the eye is viewing would be

$$\theta = \frac{h'}{d'} = \frac{1.22h}{39.5\ \text{cm}}$$

so the angular magnification achieved by viewing the image rather than the leaf directly is

$$m = \frac{\theta}{\theta_0} = \frac{1.22h/39.6\ \text{cm}}{h/197\ \text{cm}} = 1.22\left(\frac{197\ \text{cm}}{39.5\ \text{cm}}\right) = 6.08 \qquad \Diamond$$

27. The lenses of an astronomical telescope are 92 cm apart when adjusted for viewing a distant object with minimum eyestrain. The angular magnification produced by the telescope is 45. Compute the focal length of each lens.

Solution

An astronomical telescope views very distant objects, so the object distance for the objective lens is $p_o \approx \infty$. The image is formed at its focal point $(q_o = f_o)$.

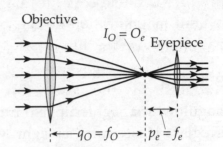

This image serves as the object for the eyepiece lens. If the telescope is adjusted for minimum eyestrain, the object for the eyepiece is located at its focal point $(p_e = f_e)$, and parallel rays enter the observer's eye. Under these conditions, the length of the telescope (that is, the distance between the objective and eyepiece lenses) is

$L = f_o + f_e$, and the angular magnification is $m = \dfrac{f_o}{f_e}$.

It is given that the lenses of the telescope are 92 cm apart. Hence,

$$L = f_o + f_e = 92 \text{ cm} \qquad\qquad [1]$$

Also, the angular magnification produced by this telescope is $m = 45$

Therefore,

$$m = \frac{f_o}{f_e} = 45, \text{ or } f_o = 45 f_e \qquad\qquad [2]$$

Substituting Equation [2] into Equation [1] gives

$$45 f_e + f_e = 46 f_e = 92 \text{ cm}$$

or the focal length of the eyepiece is $\qquad f_e = 2.0 \text{ cm} \qquad\qquad \Diamond$

Equation [2] then gives the focal length of the objective lens as

$$f_o = 45 f_e = 45(2.0 \text{ cm}) = 90 \text{ cm} \qquad\qquad \Diamond$$

31.   A person decides to use an old pair of eyeglasses to make some optical instruments. He knows that the near point in his left eye is 50.0 cm, and the near point in his right eye is 100 cm. (a) What is the maximum angular magnification he can produce in a telescope? (b) If he places the lenses 10.0 cm apart, what is the maximum overall magnification he can produce in a microscope? (To solve part (b), go back to basics and use the thin-lens equation.)

Solution

To allow a farsighted person to see objects at a normal reading distance clearly, the eyeglass lenses form upright, virtual images at the near point of each eye when the object is 25 cm in front of the lens. Thus, we use the thin lens equation to find the focal length of each lens. For the left-side lens, using $q_L = -50.0$ cm when $p_L = 25.0$ cm gives

$$f_L = \frac{p_L q_L}{p_L + q_L} = \frac{(25.0 \text{ cm})(-50.0 \text{ cm})}{25.0 \text{ cm} - 50.0 \text{ cm}} = 50.0 \text{ cm}$$

Similarly, for the right-side lens, using $q_R = -100$ cm when $p_R = 25.0$ cm gives $f_R = 33.3$ cm.

(a)   The angular magnification produced by an astronomical telescope is $m = f_{objective} / f_{eyepiece}$. To achieve maximum magnification, one should use the lens for the left eye as the objective lens and that for the right eye as the eyepiece of the telescope. This yields a magnification of

$$m = \frac{f_L}{f_R} = \frac{50.0 \text{ cm}}{33.3 \text{ cm}} = 1.50$$ ◊

(b)   For maximum magnification by the compound microscope, the eyepiece should have the shortest focal length available and should form its image as close as a normal eye can focus $(q_e = -25.0 \text{ cm})$.

Thus, we use the right-side lens as the eyepiece $(f_e = 33.3 \text{ cm})$. In order for the eyepiece to form its image 25.0 cm in front of the eye, its object distance (from the thin lens equation) must be

$$p_e = \frac{q_e f_e}{q_e - f_e} = \frac{(-25.0 \text{ cm})(33.3 \text{ cm})}{-25.0 \text{ cm} - 33.3 \text{ cm}} = 14.3 \text{ cm}$$

Since the object for the eyepiece is the image formed by the objective lens, we find that the image distance for the objective lens must be

$$q_o = L - p_e = 10.0 \text{ cm} - 14.3 \text{ cm} = -4.3 \text{ cm}$$

Therefore, the objective lens (the left-side lens from the eyeglasses) must form a virtual image 4.3 cm in front of it. This requires an object distance of

$$p_o = \frac{q_o f_o}{q_o - f_o} = \frac{(-4.3 \text{ cm})(50.0 \text{ cm})}{-4.3 \text{ cm} - 50.0 \text{ cm}} = 3.9 \text{ cm}$$

The maximum overall magnification from a 10.0 cm long compound microscope made from the available lenses is then

$$M = M_o M_e = \left(-\frac{q_o}{p_o}\right)\left(-\frac{q_e}{p_e}\right) = \left(-\frac{-4.3 \text{ cm}}{3.9 \text{ cm}}\right)\left(-\frac{-25.0 \text{ cm}}{14.3 \text{ cm}}\right) = +1.9 \qquad \lozenge$$

35.   To increase the resolving power of a microscope, the object and the objective are immersed in oil ($n = 1.5$). If the limiting angle of resolution without the oil is $0.60\ \mu\text{rad}$, what is the limiting angle of resolution with the oil? [*Hint*: The oil changes the wavelength of the light.]

Solution

Using Rayleigh's criterion, the limiting angle of resolution of a circular aperture is

$$\theta_{min} = 1.22\frac{\lambda_n}{D}$$

In this expression, D is the diameter of the aperture, and λ_n is the wavelength of the light in the medium surrounding the aperture.

When in air, the limiting angle of resolution of the microscope's objective lens is

$$\theta_{min}\big|_{air} = 1.22\frac{\lambda_{air}}{D} = 1.22\frac{\lambda}{n_{air}D}$$

where λ is the wavelength of the light when in a vacuum, and n_{air} is the index of refraction of air.

When immersed in oil, the limiting angle of resolution of this aperture is

$$\theta_{min}\big|_{oil} = 1.22\frac{\lambda_{oil}}{D} = 1.22\frac{\lambda}{n_{oil}D}$$

Forming a ratio of these results, recognizing that λ and D are the same in both cases, yields

$$\frac{\theta_{min}\big|_{oil}}{\theta_{min}\big|_{air}} = \frac{1.22\,\lambda/n_{oil}D}{1.22\,\lambda/n_{air}D} = \frac{n_{air}}{n_{oil}}$$

or $\theta_{min}\big|_{oil} = \left(\dfrac{n_{air}}{n_{oil}}\right)\theta_{min}\big|_{air} = \left(\dfrac{1.0}{1.5}\right)(0.60\ \mu\text{rad}) = 0.40\ \mu\text{rad}$ ◊

41. A 15.0-cm-long grating has 6 000 slits per centimeter. Can two lines of wavelengths 600.000 nm and 600.003 nm be separated with this grating? Explain.

Solution

The resolving power required to separate the two specified wavelengths is

$$R_{\text{required}} = \frac{\lambda}{\Delta\lambda} = \frac{600.000 \text{ nm}}{(600.003 - 600.000) \text{ nm}} = 2.00 \times 10^5$$

The resolving power of a diffraction grating in the mth order is $R = Nm$, where N is the total number of slits on the grating. The number of slits on the specified grating is

$$N = (15.0 \text{ cm})(6\ 000 \text{ slits/cm}) = 9.00 \times 10^4$$

The spacing between adjacent slits on this grating is

$$d = \frac{1.00 \text{ cm}}{6\ 000} = 1.67 \times 10^{-4} \text{ cm} = 1.67 \times 10^{-6} \text{ m}$$

Because the angle of diffraction cannot exceed 90°, the grating equation gives the maximum order of 600 nm light one could hope to observe with this grating as

$$m = \frac{d \sin\theta_{\text{max}}}{\lambda} = \frac{(1.67 \times 10^{-6} \text{ m})\sin 90°}{600 \times 10^{-9} \text{ m}} = 2.78$$

The order number must have integer values, and the third order cannot be reached. Therefore, when using this grating, the last observable order in the 600 nm region of the spectrum is the second order $(m_{\text{max}} = 2)$. The maximum resolving power in this region for the specified grating is then

$$R_{\text{max}} = Nm_{\text{max}} = (9.00 \times 10^4)(2) = 1.80 \times 10^5$$

Since this value is less than the required resolving power, it is not possible to separate the two wavelengths in question using this grating. ◊

46. The Michelson interferometer can be used to measure the index of refraction of a gas by placing an evacuated transparent tube in the light path along one arm of the device. Fringe shifts occur as the gas is slowly added to the tube. Assume that 600-nm light is used, that the tube is 5.00 cm long, and that 160 fringe shifts occur as the pressure of the gas in the tube increases to atmospheric pressure. What is the index of refraction of the gas? [*Hint:* The fringe shifts occur because the wavelength of the light changes inside the gas-filled tube.]

Solution

In the fringe pattern produced by a Michelson interferometer, a fringe shift occurs each time the effective length of one arm of the interferometer changes by a quarter-wavelength. In this case, the effective length of the arm changes because the number of wavelengths that fit between the ends of the tube changes as the tube fills with gas. Four fringe shifts occur for each addition wavelength that can be fitted within the length of the tube.

Thus, if L is the length of the tube, λ is the wavelength of the light in a vacuum, and $\lambda_n = \lambda/n_{gas}$ is the wavelength of the light when traveling in a gas having refractive index n_{gas}, the number of shifts one should observe as the tube is filling with gas is

$$N = 4\left[\frac{L}{\lambda_n} - \frac{L}{\lambda}\right] = 4\left[\frac{L}{\lambda/n_{gas}} - \frac{L}{\lambda}\right] = \frac{4L}{\lambda}\left(n_{gas} - 1\right)$$

The index of refraction of the gas used to fill the tube must then be

$$n_{gas} = 1 + \frac{N\lambda}{4L}$$

For the given situation, $L = 5.00$ cm and $\lambda = 600$ nm. Thus, if 160 fringe shifts are observed as the tube fills, the index of refraction of the gas in the tube is

$$n_{gas} = 1 + \frac{160\left(600 \times 10^{-9}\ \text{m}\right)}{4\left(5.00 \times 10^{-2}\ \text{m}\right)} = 1.000\ 5$$

◊

49. The near point of an eye is 75.0 cm. (a) What should be the power of a corrective lens prescribed to enable the eye to see an object clearly at 25.0 cm? (b) If, using the corrective lens, the person can see an object clearly at 26.0 cm, but not at 25.0 cm, by how many diopters did the lens grinder miss the prescription?

Solution

(a) If the corrective lens is to work properly, it should form an upright, virtual image at the near point of the eye $(q = -75.0 \text{ cm})$ when the object is located 25.0 cm in front of the eye $(p = +25.0 \text{ cm})$.

The thin lens equation gives the required optical power of the corrective lens as

$$\mathcal{P}_{required} = \frac{1}{f_{in \atop meters}} = \frac{1}{p} + \frac{1}{q} = \frac{1}{0.250 \text{ m}} - \frac{1}{0.750 \text{ m}} = \frac{3-1}{0.750 \text{ m}}$$

or $\quad \mathcal{P}_{required} = \dfrac{2}{0.750 \text{ m}} = +2.67 \text{ diopters}$ ◊

(b) If the user actually sees a clear image (located at the near point with $q = -75.0 \text{ cm}$) when the object is 26.0 cm in front of the eye rather than the desired 25.0 cm, the actual power of the provided corrective lens is

$$\mathcal{P}_{actual} = \frac{1}{f} = \frac{1}{p} + \frac{1}{q} = \frac{1}{0.260 \text{ m}} - \frac{1}{0.750 \text{ m}} = \frac{0.750 \text{ m} - 0.260 \text{ m}}{(0.260 \text{ m})(0.750 \text{ m})}$$

or $\quad \mathcal{P}_{actual} = \dfrac{+0.490 \text{ m}}{(0.260 \text{ m})(0.750 \text{ m})} = +2.51 \text{ diopters}$

The magnitude of the error made in grinding this corrective lens is then seen to be

$$\Delta \mathcal{P} = \mathcal{P}_{required} - \mathcal{P}_{actual} = 2.67 \text{ diopters} - 2.51 \text{ diopters} = 0.16 \text{ diopters}$$

or the supplied lens has a power that is 0.16 diopters too low. ◊

51.   A cataract-impaired lens in an eye may be surgically removed and replaced by a manufactured lens. The focal length required for the new lens is determined by the lens-to-retina distance, which is measured by a sonarlike device, and by the requirement that the implant provide for correct distance vision. (a) If the distance from lens to retina is 22.4 mm, calculate the power of the implanted lens in diopters. (b) Since there is no accommodation and the implant allows for correct distance vision, a corrective lens for close work or reading must be used. Assume a reading distance of 33.0 cm, and calculate the power of the lens in the reading glasses.

Solution

(a)   To provide correct distant vision, the implanted lens must focus a real image of very distant objects $(p \rightarrow \infty)$ on the retina with an image distance of $q = +22.4 \text{ mm}$. The thin lens equation then gives the required focal length for the implanted lens as

$$\frac{1}{f} = \frac{1}{p} + \frac{1}{q} = 0 + \frac{1}{q} \qquad \text{or} \qquad f = q = +22.4 \text{ mm}$$

The optical power of this lens will be

$$\mathscr{P}_{implant} = \frac{1}{f} = \frac{1}{+22.4 \times 10^{-3} \text{ m}} = +44.6 \text{ diopters} \qquad \Diamond$$

(b)   The implanted lens will focus parallel rays (from very distant objects) on the retina but cannot adjust to focus on nearby objects. To allow the patient to read print held 33.0 cm in front of the eye, reading glasses must be used. The reading glasses should take the divergent rays from the printed page $(p = +33.0 \text{ cm})$ and render them parallel $(q \rightarrow \infty)$ so the implanted lens in the eye can then focus them properly on the retina.

The thin lens equation gives the required optical power of the reading glasses as

$$\mathscr{P}_{reading} = \frac{1}{f_{reading}} = \frac{1}{p} + \frac{1}{q} = \frac{1}{+0.330 \text{ m}} + 0 = +3.03 \text{ diopters} \qquad \Diamond$$

55.   A laboratory (astronomical) telescope is used to view a scale that is 300 cm from the objective, which has a focal length of 20.0 cm; the eyepiece has a focal length of 2.00 cm. Calculate the angular magnification when the telescope is adjusted for minimum eyestrain. [*Note:* The object is not at infinity, so the simple expression $m = f_o / f_e$ is not sufficiently accurate for this problem. Also, assume small angles, so that $\tan\theta \approx \theta$.]

Solution

The angular magnification is $m = \theta / \theta_0$ where θ is the angle subtended by the final image, and θ_0 is the angle subtended by the original object as shown in the figure below. When the telescope is adjusted for minimum eyestrain, the rays entering the eye are parallel. Thus, the objective lens must form its image at the focal point of the eyepiece.

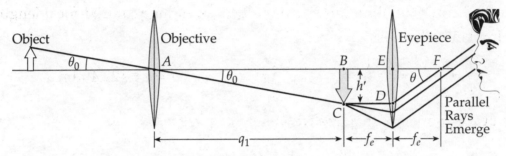

Using triangle ABC and recognizing that angle θ_0 will be very small,

we see that

$$\theta_0 \approx \tan\theta_0 = \frac{h'}{q_1}$$

where h' is the height of the image formed by the objective lens.

From triangle DEF,

$$\theta \approx \tan\theta = \frac{h'}{f_e}$$

The angular magnification is then $m = \dfrac{\theta}{\theta_0} = \dfrac{h'/f_e}{h'/q_1} = \dfrac{q_1}{f_e}$

The thin lens equation gives the image distance for the objective lens as

$$q_1 = \frac{p_1 f_1}{p_1 - f_1} = \frac{(300 \text{ cm})(20.0 \text{ cm})}{300 \text{ cm} - 20.0 \text{ cm}} = 21.4 \text{ cm}$$

so, if $f_e = 2.00 \text{ cm}$, we have $m = \dfrac{q_1}{f_e} = \dfrac{21.4 \text{ cm}}{2.00 \text{ cm}} = 10.7$ ◊

Chapter 26
Relativity

NOTES FROM SELECTED CHAPTER SECTIONS

26.2 The Principle of Galilean Relativity

According to the principle of Galilean relativity, the laws of mechanics are the same in all inertial frames of reference. **Inertial frames of reference are those coordinate systems which are at rest with respect to one another or which move at constant velocity with respect to one another. There is no preferred frame of reference for describing the laws of mechanics.**

26.4 The Michelson-Morley Experiment

This experiment was designed to detect the velocity of the Earth with respect to the hypothetical luminiferous ether and thereby confirm the prediction of an absolute frame of reference. The instrument used is called an interferometer (refer to Chapter 25, Section 25.7), and the measurement involved an attempt to observe a fringe shift in the interference pattern of two light beams. The light beams were directed along perpendicular paths, one parallel to, and one perpendicular to the "ether wind" and then combined to form the interference pattern. **The outcome of the experiment was negative,** contradicting the ether hypothesis. Light is now understood to be an electromagnetic wave that requires no medium for its propagation.

26.5 Einstein's Principle of Relativity

Einstein's special theory of relativity is based upon two postulates:

1. The laws of physics are the same in all inertial frames of reference. **An inertial frame of reference is a non-accelerated frame**.

2. The speed of light in a vacuum has the same value as measured by all observers in all inertial reference frames. **The measured value for the speed of light is independent of the motion of the observer or of the motion of**

the light source; this is consistent with the negative result of the Michelson-Morley experiment.

26.6 Consequences of Special Relativity

The concepts of length, time, and simultaneity in relativistic mechanics are quite different from those concepts in Newtonian mechanics. **In relativistic mechanics length and time are not absolute.** Following are some of the terms used to describe the consequences of special relativity:

- **Simultaneity**: Two events that are simultaneous in one reference frame are, in general, not simultaneous in another frame which is moving with respect to the first.

- **Proper time**: The time interval between two events as measured by an observer who sees the two events occur at the same position.

- **Time dilation**: A moving clock runs slower than an identical clock at rest with respect to the observer.

- **Proper length**: The length of an object measured by an observer at rest relative to the object (moving with the object).

- **Length contraction**: The length of an object measured in a reference frame that is moving with respect to the object is always less than the proper length if this length is oriented parallel to the motion. *Lengths oriented perpendicular to the motion do not exhibit length contraction.*

26.7 Relativistic Momentum

To account for relativistic effects, it is necessary to modify the classical definition of momentum to satisfy the following conditions:

- The relativistic momentum must be conserved in all inertial frames.

- The relativistic momentum must approach the classical value, mv, as the ratio (v/c) approaches zero.

26.8 Relativistic Addition of Velocities
26.9 Relativistic Energy and The Equivalence of Mass and Energy

In order to be compatible with the principles of relativity, the expression for addition of relativistic velocities includes a correction to Galilean relativity based on length contraction and time dilation.

A particle has a rest energy proportional to the value of its inertial mass, $E_R = mc^2$. When the particle is in motion, it has a total energy equal to the sum of its kinetic energy and its rest energy.

26.10 Pair Production and Annihilation

Pair production is a process in which the energy of a photon is converted into mass. In this process, a particle and its antiparticle are simultaneously produced, while the photon disappears. The minimum energy that a photon must have to produce a particle-antiparticle pair can be found using conservation of energy by equating the photon energy to the total rest energy of the pair.

Pair production cannot occur in a vacuum but can only take place in the presence of a large atomic nucleus. **The heavy nucleus must participate in the interaction in order that energy and momentum be conserved simultaneously.**

Pair annihilation is a process in which an particle-antiparticle pair combine with each other, disappear, and produce two photons, moving in opposite directions, each with the same energy and magnitude of momentum.

26.11 General Relativity

Einstein's postulates of general relativity are:

*   All the laws of nature have the same form for observers in any reference frame (inertial or non-inertial).

*   In the vicinity of any given point, a gravitational field is equivalent to an accelerated frame of reference without a gravitational field (the principle of equivalence).

EQUATIONS AND CONCEPTS

Important consequences of the theory of special relativity:

- **Time dilation** — A time interval Δt measured by an observer moving with respect to a clock is longer than the time interval Δt_p (the proper time) measured by an observer at rest with respect to the clock. *Moving clocks run slower than clocks at rest with respect to an observer.*

$$\Delta t = \frac{\Delta t_p}{\sqrt{1 - v^2/c^2}} = \gamma \, \Delta t_p \qquad (26.7)$$

$$\gamma = \frac{1}{\sqrt{1 - v^2/c^2}} \qquad (26.8)$$

- **Length contraction** — If an object has a length L_p (the proper length) when measured by an observer at rest with respect to the object, the length L measured by an observer as the object moves along a direction parallel to its length will be less than L_p. *Length contraction occurs only along the direction of motion.*

$$L = \frac{L_p}{\gamma} = L_p \sqrt{1 - \frac{v^2}{c^2}} \qquad (26.9)$$

- **Simultaneity** — events observed as simultaneous in one frame of reference are not necessarily observed as simultaneous in a second frame moving relative to the first.

Some important equations of relativistic mechanics:

- The **relativistic linear momentum** of a particle with mass m and moving with speed v satisfies the following conditions:

  (1) Momentum is conserved in all collisions.

  (2) The relativistic value of momentum approaches the classical value (mv) as v approaches zero.

$$p \equiv \frac{mv}{\sqrt{1 - v^2/c^2}} = \gamma mv \qquad (26.10)$$

- The **relativistic kinetic energy** of a particle of mass m moving with a speed v, includes the rest energy term mc^2.

$$KE = \gamma mc^2 - mc^2 \qquad (26.12)$$

- The **rest energy** of a particle is independent of the speed of the particle. The mass m must have the same value in all inertial frames.

$$E_R = mc^2 \qquad (26.13)$$

- The **total energy** E of a particle is the sum of the kinetic energy and the rest energy. *This expression shows that mass is a form of energy. Equation 26.16 is useful in cases where the speed of an object is not known.*

$$E = KE + mc^2 = \gamma mc^2 \qquad (26.14)$$

$$E = \frac{mc^2}{\sqrt{1 - v^2/c^2}} \qquad (26.15)$$

$$E^2 = p^2c^2 + (mc^2)^2 \qquad (26.16)$$

- **Relativistic addition of velocities.** *Carefully observe the pattern of the subscripts of the velocities.*

$$v_{ab} = \frac{v_{ad} + v_{db}}{1 + \dfrac{v_{ad}v_{db}}{c^2}} \qquad (26.11)$$

v_{ad} = speed of object "a" relative to moving frame d

v_{db} = speed of moving frame d relative to moving frame b

v_{ab} = speed of object "a" relative to moving frame b

Photons ($m = 0$) travel with the speed of light. Equation 26.17 is an exact expression relating energy and momentum for particles which have zero mass.

$$E = pc \qquad (26.17)$$

The **electron volt** (eV) is a convenient energy unit to use to express the energies of electrons and other subatomic particles.

$$1 \text{ eV} = 1.60 \times 10^{-19} \text{ J}$$
$$m_e c^2 = 0.511 \text{ MeV}$$
$$= 8.20 \times 10^{-14} \text{ J}$$

The **minimum photon frequency for electron-positron pair production** can be found from conservation of energy.

$$hf_{min} = 2m_e c^2 \qquad (26.18)$$

REVIEW CHECKLIST

▷ State Einstein's two postulates of the special theory of relativity.

▷ Understand the Michelson-Morley experiment, its objectives, results, and the significance of its outcome.

▷ Understand the idea of simultaneity and the fact that simultaneity is not an absolute concept. That is, two events which are simultaneous in one reference frame are not simultaneous when viewed from a second frame moving with respect to the first.

▷ Make calculations using the equations for time dilation and length contraction.

▷ State the correct relativistic expressions for the momentum, kinetic energy, and total energy of a particle. Make calculations using these equations.

SOLUTIONS TO SELECTED END-OF-CHAPTER PROBLEMS

7. The average lifetime of a pi meson in its own frame of reference (that is, the proper lifetime) is 2.6×10^{-8} s. If the meson moves with a speed of $0.98\,c$, what is (a) its mean lifetime as measured by an observer on Earth, and (b) the average distance it travels before decaying, as measured by an observer on Earth? (c) What distance would it travel if time dilation did not occur?

Solution

(a) If the proper lifetime of the pi meson is $\Delta t_p = 2.6 \times 10^{-8}$ s, the lifetime measured by an observer on Earth (moving at speed $v = 0.98\,c$ relative to the meson) is

$$\Delta t = \gamma \left(\Delta t_p \right) = \frac{\Delta t_p}{\sqrt{1 - (v/c)^2}} = \frac{2.6 \times 10^{-8} \text{ s}}{\sqrt{1 - (0.98)^2}} = 1.3 \times 10^{-7} \text{ s} \qquad \lozenge$$

(b) The observer on Earth sees the meson travel at $0.98c$ for $\Delta t = 1.3 \times 10^{-7}$ s. Thus, the distance the meson travels during its lifetime according to this observer is

$$L_p = v \left(\Delta t \right) = 0.98 \left(3.0 \times 10^8 \text{ m/s} \right) \left(1.3 \times 10^{-7} \text{ s} \right) = 38 \text{ m} \qquad \lozenge$$

(c) If time dilation did not occur, the Earth-based observer would have seen the meson travel at $0.98c$ for its proper lifetime of $\Delta t_p = 2.6 \times 10^{-8}$ s. Thus, the distance this observer would see the meson travel under these circumstances would be

$$L = v \left(\Delta t_p \right) = 0.98 \left(3.0 \times 10^8 \text{ m/s} \right) \left(2.6 \times 10^{-8} \text{ s} \right) = 7.6 \text{ m} \qquad \lozenge$$

11.   The proper length of one spaceship is three times that of another. The two spaceships are traveling in the same direction and, while both are passing overhead, an Earth observer measures the two spaceships to have the same length. If the slower spaceship is moving with a speed of 0.350c, determine the speed of the faster spaceship.

Solution

The faster object is the one that will appear to be contracted the most to the Earth-based observer. Since this observer measures the two moving ships to have the same length, the faster one must have the greater proper length, or $L_{pf} = 3L_{ps}$.

The contracted lengths that the Earth-based observer measures for the two ships are:

$$L_f = \frac{L_{pf}}{\gamma_f} = L_{pf}\sqrt{1-\left(v_f/c\right)^2}$$

and
$$L_s = \frac{L_{ps}}{\gamma_s} = L_{ps}\sqrt{1-\left(v_s/c\right)^2} = L_{ps}\sqrt{1-(0.350)^2} = 0.937L_{ps}$$

But, since this observer determines that $L_f = L_s$, and we have already determined that $L_{pf} = 3L_{ps}$, we find that

$$\left(3L_{ps}\right)\sqrt{1-\left(v_f/c\right)^2} = 0.937L_{ps} \qquad \text{or} \qquad \sqrt{1-\left(v_f/c\right)^2} = 0.312$$

Thus, $v_f = c\sqrt{1-(0.312)^3} = 0.950\,c$ ◊

16. An interstellar space probe is launched from Earth. After a brief period of acceleration it moves with a constant velocity, 70.0% of the speed of light. Its nuclear-powered batteries supply the energy to keep its data transmitter active continuously. The batteries have a lifetime of 15.0 years as measured in a rest frame. (a) How long do the batteries on the space probe last as measured by mission control on Earth? (b) How far is the probe from Earth when its batteries fail, as measured by mission control? (c) How far is the probe from Earth, as measured by its built-in trip odometer, when its batteries fail? (d) For what total time after launch is data received from the probe by mission control? Note that radio waves travel at the speed of light and fill the space between the probe and Earth at the time the battery fails.

Solution

(a) The batteries have a lifetime of $\Delta t_p = 15.0$ yr in their own rest frame (the reference frame of the ship). Observers on Earth are moving at speed $v = 0.700\,c$ relative to this reference frame and hence measure a dilated lifetime of

$$\Delta t = \gamma\left(\Delta t_p\right) = \frac{\Delta t_p}{\sqrt{1-(v/c)^2}} = \frac{15.0 \text{ yr}}{\sqrt{1-(0.700)^2}} = 21.0 \text{ yr} \qquad \Diamond$$

(b) Mission control sees the ship travel outward at $v = 0.700\,c$ for 21.0 yr before battery failure. Thus, the distance they measure it to have traveled is

$$L_p = v(\Delta t) = (0.700\,c)(21.0 \text{ yr}) = (14.7 \text{ yr})c = 14.7 \text{ ly} \qquad \Diamond$$

(c) The probe sees Earth recede at $v = 0.700\,c$ for $\Delta t_p = 15.0$ yr (as measured by the probe's clock) before battery failure. The reading on the probe's trip odometer is

$$L = v\left(\Delta t_p\right) = (0.700\,c)(15.0 \text{ yr}) = (10.5 \text{ yr})c = 10.5 \text{ ly} \qquad \Diamond$$

(d) According to the Earth-based clock, mission control receives signals for 21.0 yr before the batteries fail. After the battery fails, mission control continues to receive signals until the information that left the ship at the instant the battery failed has traveled 14.7 ly (traveling at speed c) back to Earth. Thus, they receive signals for a total time of

$$\Delta t_{total} = \Delta t + (14.7 \text{ ly}/c) = 21.0 \text{ yr} + 14.7 \text{ yr} = 35.7 \text{ yr} \qquad \Diamond$$

19.   An unstable particle at rest breaks up into two fragments of *unequal mass*. The mass of the lighter fragment is 2.50×10^{-28} kg and that of the heavier fragment is 1.67×10^{-27} kg. If the lighter fragment has a speed of 0.893c after the breakup, what is the speed of the heavier fragment?

Solution

The unstable particle is initially at rest, so the total momentum is zero before breakup. To conserve momentum, the total momentum after breakup must also be zero, meaning that the momenta of the two fragments must have equal magnitudes and opposite directions. If we label the lighter fragment 1 and the heavier one 2, equating magnitudes of momenta gives

$$p_2 = p_1 = \gamma_1 m_1 v_1 = \frac{\left(2.50 \times 10^{-28} \text{ kg}\right)\left(0.893c\right)}{\sqrt{1 - \left(0.893\right)^2}} = \left(4.96 \times 10^{-28} \text{ kg}\right)c$$

We also know that $p_2 = \gamma_2 m_2 v_2$

so $\dfrac{\left(1.67 \times 10^{-27} \text{ kg}\right)v_2}{\sqrt{1 - \left(v_2/c\right)^2}} = \left(4.96 \times 10^{-28} \text{ kg}\right)c$

This reduces to $3.37\left(v_2/c\right) = \sqrt{1 - \left(v_2/c\right)^2}$

Squaring both sides and solving for the speed of the heavier fragment gives

$$v_2 = 0.285c \qquad\qquad \Diamond$$

23. A Klingon spaceship moves away from Earth at a speed of 0.800c (Fig. P26.23). The starship *Enterprise* pursues at a speed of 0.900c relative to Earth. Observers on Earth see the *Enterprise* overtaking the Klingon ship at a relative speed of 0.100c. With what speed is the *Enterprise* overtaking the Klingon ship as seen by the crew of the *Enterprise*?

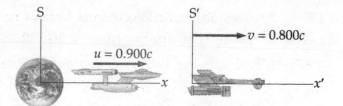

Figure P26.23

Solution

We take the direction of both ships motion relative to Earth to be the positive direction. Then, the velocity of the Klingon ship relative to Observers on Earth is $v_{KO} = +0.800\,c$, and the velocity of the *Enterprise* relative to Observers on Earth is $v_{EO} = +0.900\,c$.

The relativistic velocity addition equation gives the velocity of the *Enterprise* relative to the Klingon ship as

$$v_{EK} = \frac{v_{EO} + v_{OK}}{1 + \dfrac{v_{EO} \cdot v_{OK}}{c^2}}$$

Carefully observe the pattern of the subscripts in this expression. Note that in the numerator, the leftmost subscript of the first term and the rightmost subscript of the second term together form the subscript of the quantity being calculated on the left side of the equation. Also, in the second term of the denominator, the leftmost subscript of the first factor and the rightmost subscript of the second factor together form the subscript of the quantity on the left of the equation. In both the numerator and the denominator, the inner subscripts are the same and refer to the intermediate reference frame used in the calculation.

In the velocity addition equation above, the velocity v_{OK} is the velocity of the Earth-based observers relative to the Klingon ship. This velocity is given by

$$v_{OK} = -v_{KO} = -(+0.800\,c) = -0.800\,c$$

Thus,

$$v_{EK} = \frac{v_{EO} + v_{OK}}{1 + \dfrac{v_{EO} \cdot v_{OK}}{c^2}} = \frac{+0.900\,c + (-0.800\,c)}{1 + \dfrac{(+0.900\,c)(-0.800\,c)}{c^2}} = \frac{0.100\,c}{1 - 0.720} = +0.357\,c \qquad \Diamond$$

27. Spaceship I, which contains students taking a physics exam, approaches Earth with a speed of 0.60c, while spaceship II, which contains instructors proctoring the exam, moves away from Earth at 0.28c, as in Figure P26.27. If the instructors in spaceship II stop the exam after 50 min have passed *on their clock*, how long does the exam last as measured by (a) the students? (b) an observer on Earth?

Figure P26.27

Solution

We take toward the right in Figure P26.27 to be the positive direction. Then the velocity of the students on ship I relative to Earth is $v_{SE} = +0.60\,c$, the velocity of the instructors on ship II relative to Earth is $v_{IE} = +0.28\,c$, and the velocity of Earth relative to the instructor's ship is $v_{EI} = -v_{IE}$.

(a)   The velocity of the students relative to the instructors is given by the relativistic velocity addition equation as

$$v_{SI} = \frac{v_{SE} + v_{EI}}{1 + \dfrac{v_{SE} \cdot v_{EI}}{c^2}} = \frac{v_{SE} + (-v_{IE})}{1 + \dfrac{v_{SE} \cdot (-v_{IE})}{c^2}} = \frac{+0.60\,c - 0.28\,c}{1 + \dfrac{(0.60\,c)(-0.28\,c)}{c^2}} = +0.38\,c$$

Thus, as measured by the students, the duration of the exam Δt_S (which has a proper time interval of $\Delta t_p = 50$ min measured by the instructors on their clock) is

$$\Delta t_S = \gamma_{SI}\left(\Delta t_p\right) = \frac{\Delta t_p}{\sqrt{1 - \left(v_{SI}/c\right)^2}} = \frac{50 \text{ min}}{\sqrt{1 - (0.38)^2}} = 54 \text{ min} \qquad \Diamond$$

(b)   The relative velocity between the instructors and Earth is $v_{IE} = +0.28\,c$, so the duration of the exam Δt_E (having a proper time of 50 min in the instructor's reference frame) as measured by observers on Earth is

$$\Delta t_E = \gamma_{EI}\left(\Delta t_p\right) = \frac{\Delta t_p}{\sqrt{1 - \left(v_{EI}/c\right)^2}} = \frac{50 \text{ min}}{\sqrt{1 - (-0.28)^2}} = 52 \text{ min} \qquad \Diamond$$

33.   A cube of steel has a volume of 1.00 cm^3 and a mass of 8.00 g when at rest on Earth. If this cube is now given a speed $v = 0.900c$, what is its density as measured by a stationary observer? Note that relativistic density is E_R/c^2V.

Solution

Assume that the cube moves in a direction parallel to one of its edges. The dimension of the cube parallel to the motion will undergo length contraction, while the other two dimensions (being perpendicular to the motion) will be unchanged. If L_p is the proper length of each dimension of the cube (measured in the rest frame of the moving cube), the volume measured by a stationary observer will be

$$V = \left(L_p\right)\left(L_p\right)\left(L_p\sqrt{1-(v/c)^2}\right) = \left(L_p\right)^3\sqrt{1-(v/c)^2} = V_p\sqrt{1-(v/c)^2}$$

The density measured by the stationary observer is then

$$\rho = \frac{m}{V} = \frac{m}{V_p\sqrt{1-(v/c)^2}} = \frac{8.00\ \text{g}}{\left(1.00\ \text{cm}^3\right)\sqrt{1-(0.900)^2}} = 18.4\ \text{g/cm}^3 \qquad \lozenge$$

Note that this expression for the relativistic density can be written as

$$\rho = \frac{m}{V} = \frac{mc^2}{c^2V} = \frac{E_R}{c^2V}$$

as suggested in the statement of the problem.

37.   Two photons are produced when a proton and an antiproton annihilate each other. What are the minimum frequency and the corresponding wavelength of each photon?

Solution

The energy of a photon is $E_\gamma = hf$

and its momentum is $p_\gamma = \dfrac{E_\gamma}{c}$

where h is Planck's constant, c is the speed of light in a vacuum, and f is the frequency of the photon.

In pair annihilation, the total energy of the annihilating particles is transformed into photon energy. If this total photon energy is to be a minimum, both particles of the pair must be at rest (that is, have zero kinetic energy) before annihilation. Then, the total momentum is zero both before and after the annihilation process. If this annihilation produces two photons, the momenta of these photons must have equal magnitudes and opposite directions.

Since the two photons have equal magnitudes of momentum, the photon energies given by $E_\gamma = p_\gamma c$ are also equal. Thus, conservation of energy during the annihilation process gives

$$2E_\gamma = E_{\text{proton}} + E_{\text{antiproton}} = (KE + E_R) + (KE + E_R) = 2(0 + E_R) = 2E_R$$

or $E_\gamma = E_R$

where $E_R = m_p c^2 = (1.67 \times 10^{-27} \text{ kg})(3.00 \times 10^8 \text{ m/s})^2 = 1.50 \times 10^{-10}$ J is the rest energy of either a proton or an antiproton. The minimum frequency of each photon in a proton-antiproton annihilation is then

$$f = \frac{E_\gamma}{h} = \frac{E_R}{h} = \frac{1.50 \times 10^{-10} \text{ J}}{6.63 \times 10^{-34} \text{ J·s}} = 2.27 \times 10^{23} \text{ Hz} \qquad \lozenge$$

and the wavelength is $\lambda = \dfrac{c}{f} = \dfrac{3.00 \times 10^8 \text{ m/s}}{2.27 \times 10^{23} \text{ Hz}} = 1.32 \times 10^{-15} \text{ m} = 1.32 \text{ fm} \qquad \lozenge$

41.   What is the momentum (in units of MeV/c) of an electron with a kinetic energy of 1.00 MeV?

Solution

The relativistic expression for kinetic energy is

$$KE = E - E_R$$

where E is the total energy of the particle, and $E_R = mc^2$ is its rest energy.

For an electron, the rest energy is

$$E_R = m_e c^2 = \left(9.11 \times 10^{-31} \text{ kg}\right)\left(3.00 \times 10^8 \text{ m/s}\right)^2 \left(\frac{1 \text{ MeV}}{1.60 \times 10^{-13} \text{ J}}\right) = 0.512 \text{ MeV}$$

Thus, the total energy of an electron having a kinetic energy of 1.00 MeV is

$$E = KE + E_R = 1.00 \text{ MeV} + 0.512 \text{ MeV} = 1.51 \text{ MeV}$$

The relativistic expression relating energy and momentum, $E^2 = (pc)^2 + E_R^2$, then gives the momentum of the electron as

$$p = \frac{\sqrt{E^2 - E_R^2}}{c} = \frac{\sqrt{(1.51 \text{ MeV})^2 - (0.512 \text{ MeV})^2}}{c} = 1.42 \text{ MeV}/c \qquad \Diamond$$

45.   A radioactive nucleus moves with a speed v relative to a laboratory observer. The nucleus emits an electron in the positive x-direction with a speed of $0.70c$ relative to the decaying nucleus and a speed of $0.85c$ in the $+x$-direction relative to the laboratory observer. What is the value of v?

Solution

With the following definitions

$$v_{nL} = v = \text{velocity of the nucleus relative to the laboratory observer},$$

$$v_{eL} = +0.85c = \text{velocity of the electron relative to the laboratory observer},$$

$$v_{en} = +0.70c = \text{velocity of the electron relative to the nucleus},$$

and $v_{ne} = -v_{en} = -0.70c = \text{velocity of the nucleus relative to the electron},$

the relativistic velocity addition equation gives

$$v = v_{nL} = \frac{v_{ne} + v_{eL}}{1 + \dfrac{v_{ne} \cdot v_{eL}}{c^2}} = \frac{-0.70c + 0.85c}{1 + \dfrac{(-0.70c)(0.85c)}{c^2}} = +0.37c$$

Thus, the radioactive nucleus moves in the positive x direction at speed $0.37c$ relative to the laboratory observer.   ◊

Carefully observe the pattern of the subscripts in the relativistic velocity addition equation above. Note that these subscripts fit the pattern discussed in the solution of Problem 23 given earlier.

51.   An alien spaceship traveling 0.600*c* toward Earth launches a landing craft with an advance guard of purchasing agents. The lander travels in the same direction with a velocity of 0.800*c* relative to the spaceship. As observed on Earth, the spaceship is 0.200 lightyear from Earth when the lander is launched. (a) With what velocity is the lander observed to be approaching by observers on Earth? (b) What is the distance to Earth at the time of lander launch, as observed by the aliens on the mother ship? (c) How long does it take the lander to reach Earth, as observed by the aliens on the mother ship? (d) If the lander has a mass of 4.00×10^5, what is its kinetic energy as observed in Earth's reference frame?

Solution

(a)   Choosing toward Earth as the positive direction,

we have $v_{SE} = +0.600c =$ velocity of the ship relative to Earth ,

$v_{LS} = +0.800c =$ velocity of the lander relative to the ship ,

and $v_{LE} =$ velocity of the lander relative to Earth .

The relativistic velocity addition equation then gives

$$v_{LE} = \frac{v_{LS} + v_{SE}}{1 + \frac{(v_{LS} \cdot v_{SE})}{c^2}} = \frac{0.800c + 0.600c}{1 + \frac{(0.800c) \cdot (0.600c)}{c^2}} = +0.946c \qquad \Diamond$$

(b)   The aliens consider themselves to be at rest and see Earth (and the proper length $L_p = 0.200$ ly measured by observers on Earth) moving toward them at 0.600 *c*.

Thus, their measure of this distance is length contracted and has a value of

$$L = L_p \sqrt{1 - (v_{SE}/c)^2} = (0.200 \text{ ly})\sqrt{1 - (0.600)^2} = 0.160 \text{ ly} \qquad \Diamond$$

(c)   The aliens on the mother ship see a distance $L = 0.160$ ly initially separating the lander and Earth. From their viewpoint, the lander decreases this distance from one end $\left(\text{at a rate } |v_{LS}| = 0.800c\right)$ as it moves toward Earth, and simultaneously, Earth decreases this distance from the other end as it moves toward the lander at a rate $|v_{ES}| = |-v_{SE}| = 0.600c$. Thus, the time they will measure for this initial distance to shrink to zero (that is, for lander and Earth to meet) is

$$\Delta t = \frac{L}{|v_{LS}| + |v_{ES}|} = \frac{0.160 \text{ ly}}{0.800c + 0.600c}\left[\frac{(1 \text{ yr})c}{1 \text{ ly}}\right] = 0.114 \text{ yr} \qquad \Diamond$$

(d)   The relativistic expression for the kinetic of an object moving at speed v is

$$KE = E - E_R = \gamma E_R - E_R = (\gamma - 1)E_R = (\gamma - 1)mc^2$$

where m is the mass of the object, and $\gamma = 1 \big/ \sqrt{1 - (v/c)^2}$.

Therefore, with $v_{LE} = 0.946c$, the kinetic energy of the lander in Earth's reference frame is

$$KE = \left(\frac{1}{\sqrt{1 - (0.946)^2}} - 1 \right)(4.00 \times 10^5 \text{ kg})(3.00 \times 10^8)^2 = 7.50 \times 10^{22} \text{ J} \qquad \Diamond$$

57. A rod of length L_0 moves with a speed of v along the horizontal direction. The rod makes an angle of θ_0 with respect to the axis of a coordinate system moving with the rod. (a) Show that the length of the rod as measured by a stationary observer is given by

$$L = L_0 \left[1 - \left(\frac{v^2}{c^2} \right) \cos \theta_0 \right]^{1/2}$$

(b) Show that the angle the rod makes with the axis, as seen by the stationary observer, is given by the expression $\tan \theta = \gamma \tan \theta_0$. These results demonstrate that the rod is both contracted and rotated. (Take the lower end of the rod to be at the origin of the moving coordinate system.)

Solution

(a) The sketch at the right shows the orientation of the rod in its rest frame, which moves at speed v parallel to the x axis of the stationary observer's reference frame. The proper length of the rod (the length measured in its own rest frame) is L_0. The x- and y-components of the rod in this reference frame are

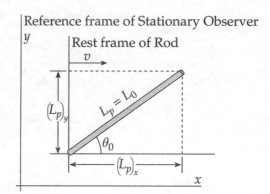

$$\left(L_p \right)_x = L_0 \cos \theta_0 \quad \text{and} \quad \left(L_p \right)_y = L_0 \sin \theta_0$$

The stationary observer sees the component parallel to the direction of motion (the x-component) length contracted while the y-component, being perpendicular to the motion, is not altered. Thus, the components of the rod's length measured by the stationary observer are

$$L_x = \left(L_p \right)_x \sqrt{1 - \left(v/c \right)^2} = L_0 \cos \theta_0 \sqrt{1 - \left(v/c \right)^2} \quad \text{and} \quad L_y = \left(L_p \right)_y = L_0 \sin \theta_0$$

The total length of the rod as measured by the stationary observer is then

$$L = \sqrt{L_x^2 + L_y^2} = L_0 \sqrt{\cos^2 \theta_0 - \left(\frac{v}{c} \right)^2 \cos^2 \theta_0 + \sin^2 \theta_0}$$

or $L = L_0 \sqrt{\left(\cos^2 \theta_0 + \sin^2 \theta_0 \right) - \left(\frac{v}{c} \right)^2 \cos^2 \theta_0} = L_0 \left[1 - \left(\frac{v}{c} \right)^2 \cos^2 \theta_0 \right]^{1/2}$ ◊

(b)   The angle the rod makes with the horizontal axis as seen by the stationary observer is found from

$$\tan\theta = \frac{L_y}{L_x} = \frac{L_0 \sin\theta_0}{L_0 \cos\theta_0 \sqrt{1-(v/c)^2}} = \left(\frac{1}{\sqrt{1-(v/c)^2}}\right)\frac{\sin\theta_0}{\cos\theta_0}$$

or $\tan\theta = \gamma \tan\theta_0$ ◊

Chapter 27

Quantum Physics

NOTES FROM SELECTED CHAPTER SECTIONS

27.1 Blackbody Radiation and Planck's Hypothesis

A **black body** is an ideal system that absorbs all radiant energy incident on it. The nature of the radiation emitted by a black body depends only on its temperature. The spectral distribution of blackbody radiation (thermal radiation) at various temperatures is sketched in the figure below.

As the temperature increases, the total radiation emitted (area under the curve) increases, while the peak of the distribution shifts to shorter wavelengths. Stefan's law describes the total power radiated as a function of temperature. The shift to shorter wavelengths is consistent with the **Wien displacement law (Equation 27.1)**. Classical theories failed to explain the spectrum of blackbody radiation as observed experimentally. An empirical formula, proposed by Max Planck, is consistent with the observed distribution at all wavelengths. Planck made two basic assumptions:

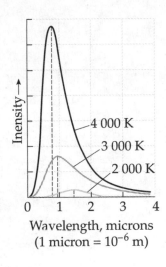

Wavelength, microns
(1 micron = 10^{-6} m)

(1) Blackbody radiation is caused by submicroscopic charged oscillators called **resonators.**

(2) The resonators could only have **discrete energies (quantum states)** given by $E_n = nhf$, where f is the oscillator frequency, n is a quantum number ($n = 1, 2, 3, \ldots$), and h is Planck's constant.

Subsequent developments showed that the quantum concept was necessary in order to explain several phenomena at the atomic level, including the photoelectric effect, the Compton effect, and atomic spectra.

27.2   The Photoelectric Effect and the Particle Theory of Light

When light, of sufficiently high frequency, is incident on certain metallic surfaces, electrons will be emitted from the surfaces. This is called the **photoelectric effect**, discovered by Hertz.

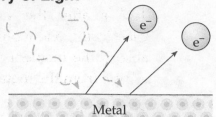

Several features of the photoelectric effect cannot be explained with classical physics or with the wave theory of light. However, each of these features can be explained and understood on the basis of the photon theory of light. These features and their explanations include:

1. **No electrons are emitted if the frequency of the incident light is below a value called the cutoff frequency, f_c.** The value of the cutoff frequency depends on the particular material being illuminated. For example, in the case of sodium, $f_c = 5.50 \times 10^{14}$ Hz.

    This observation is explained based on the quantum model. In this model, the energy of electromagnetic radiation is assumed to occur in packets called photons, each having energy $E = hf$ where $h = 6.626 \times 10^{-34}$ J·s is known as Planck's constant and, f is the frequency of the radiation. Photons in the incident light interact with individual electrons. **If the energy of an incoming photon is not equal to or greater than the work function, ϕ, of the photosensitive surface, no electrons will be ejected from the surface, regardless of the intensity of the light.**

2. If the light frequency exceeds the cutoff frequency, a photoelectric effect is observed, and **the number of photoelectrons emitted is proportional to the light intensity.** However, the **maximum kinetic energy of the photoelectrons is independent of light intensity**; a fact that cannot be explained by the concepts of classical physics.

    The fact that KE_{max} is independent of the light intensity can be understood with the following argument. If the light intensity is doubled, the number of photons is doubled, which doubles the number of photoelectrons emitted. However, their kinetic energy, which equals $hf - \phi$, depends only on the light frequency and the work function, not on the light intensity.

3. **The maximum kinetic energy of the photoelectrons increases with increasing light frequency.** The fact that KE_{max} increases with increasing frequency is easily understood from Equation 27.6, $KE_{max} = hf - \phi$.

4.   **Electrons are emitted from the surface almost instantaneously** (less than 10^{-9} s after the surface is illuminated), even at low light intensities. Classically, one would expect that the electrons would require some time to absorb the incident radiation before they acquire enough kinetic energy to escape from the metal.

The fact that the electrons are emitted almost instantaneously is consistent with the photon theory of light, in which the incident energy appears as individual photons, and there is a one-to-one interaction between photons and electrons.

27.3   X-Rays

X-Rays are a part of the electromagnetic spectrum, characterized by frequencies higher than those of ultraviolet radiation with wavelengths in the 10 to 10^{-4} nm range. They are produced when a metal (and in some cases non-metal) target is struck by high speed electrons.

The spectrum of radiation emitted by an X-ray tube has two distinct components:

Bremsstrahlung — a continuous broad spectrum of high-energy photons produced by accelerating electrons as they pass close to large positive nuclei in the metal target .

Characteristic lines — a series of sharp lines characteristic of the target material which appear superimposed on the continuous spectrum as the accelerating voltage exceeds the threshold value. These lines represent photons emitted as electrons in atoms of the target material undergo rearrangements following collisions with the incoming electrons.

27.5   The Compton Effect

The Compton effect involves the scattering of a photon by an electron. **The scattered photon undergoes a change in wavelength $\Delta\lambda$, called the Compton shift.** An increase in wavelength for a scattered photon can be explained on the basis of a scattering event in which a **portion of the energy of an incident ("bombarding") photon is transferred to an electron ("target").** The total energy and the total momentum of the photon-electron pair are each conserved during the scattering event

Using this model the predicted value of the wavelength shift is in excellent agreement with experimental results.

The figure at the right shows the Compton shift for X-rays scattered at 90^0 from graphite. In this case, the Compton shift is $\Delta\lambda = 0.002\ 36\ nm$, and λ_0 is the wavelength of photons in the incident x-ray beam.

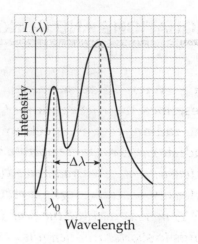

27.6 The Dual Nature of Light and Matter

The results of some experiments (i.e. photoelectric emission of electrons and Compton scattering) can only be explained on the basis of the photon model of light. Other phenomena (i.e. interference and diffraction of light) are consistent only with the wave model. **The photon theory and the wave theory complement each other—light exhibits both wave and particle characteristics.**

De Broglie postulated that a particle in motion has wave properties and a corresponding wavelength inversely proportional to the particle's momentum. The wave nature of matter was first confirmed (Davisson and Germer) by scattering electrons from crystal targets. The operation of the electron microscope is based on the wave properties of electrons.

27.7 The Wave Function

In the Schrodinger description of the manner in which matter waves change in space and time, each particle is represented by a wave function. *The probability of finding a particle at a specific location at some instant is given by evaluating the square of the wave function at that location and time.*

27.8 The Uncertainty Principle

Quantum theory predicts that **it is impossible to make exact simultaneous measurements of a particle's position and linear momentum or to measure exactly the energy of a particle in an arbitrarily small time interval**. These limitations are described by Equations 27.16 and 27.17.

EQUATIONS AND CONCEPTS

The **Wien displacement law** accurately describes the distribution of wavelengths in the radiant energy spectrum of a blackbody radiator. *As the temperature increases, the intensity of the radiation (area under the intensity vs. wavelength curve shown in Section 27.1) increases, while the peak of the distribution shifts to shorter wavelengths.*

$$\lambda_{max} T = 0.289\,8 \times 10^{-2} \text{ m} \cdot \text{K} \qquad (27.1)$$

Discrete energy values of an atomic oscillator are determined by a quantum number, n. Each discrete energy value corresponds to a quantum state. **Planck's constant** is a fundamental constant of nature.

$$E_n = nhf \qquad (27.2)$$
$$(n = 1, 2, 3, \ldots)$$

$$h = 6.626 \times 10^{-34} \text{ J} \cdot \text{s} \qquad (27.3)$$

The **energy of a quantum** or photon corresponds to the energy difference between initial and final quantum states. *An oscillator emits or absorbs energy only when there is a transition between quantum states.*

$$E = hf \qquad (27.5)$$

The **maximum kinetic energy of an ejected photoelectron** (typically a few eV) can be expressed in terms of the stopping potential or in terms of the work function of the metal, ϕ, and the frequency of the incident photon.

$$KE_{max} = e\Delta V_s \qquad (27.4)$$

$$KE_{max} = hf - \phi \qquad (27.6)$$

The **cut-off wavelength** is the maximum wavelength (corresponding to the minimum frequency) of an incident photon that will result in photoemission regardless of the intensity of the incident light.

$$\lambda_c = \frac{hc}{\phi} \qquad (27.7)$$

The **minimum wavelength X-ray** radiation, for a given accelerating voltage, is produced when an electron is completely stopped in a single collision.

$$\lambda_{min} = \frac{hc}{e\Delta V} \tag{27.9}$$

Bragg's law states the condition for constructive interference of x-rays diffracted by a crystal. The crystal planes are separated by a distance d; x-rays are incident at an angle θ to the crystal plane and are reflected at the same angle.

$$2d\sin\theta = m\lambda \tag{27.10}$$

$$(m = 1, 2, 3, \ldots)$$

The **Compton shift** is the change in wavelength of an x-ray when scattered from an electron. *The scattered x-ray makes an angle θ with the direction of the incident x-ray.*

$$\Delta\lambda = \lambda - \lambda_o = \frac{h}{m_e c}(1 - \cos\theta) \tag{27.11}$$

Compton wavelength $= h/(m_e c)$
$$= 0.002\,43 \text{ nm}$$

The **momentum and wavelength** of a photon are related as shown in Equation 27.13. This equation is obtained by combining two equations involving the energy of a photon: $p = E/c$ and $E = hc/\lambda$.

$$p = \frac{E}{c} = \frac{h}{\lambda} \tag{27.13}$$

The **de Broglie wavelength** of a particle is inversely proportional to the momentum of the particle. *The waves associated with material particles are called **matter waves** with frequencies given by Equation 27.15.*

$$\lambda = \frac{h}{mv} \tag{27.14}$$

$$f = \frac{E}{h} \tag{27.15}$$

The **Heisenberg uncertainty principle** can be stated in two forms:

- Simultaneous measurements of **position and momentum** with respective uncertainties Δx and Δp_x.

$$\Delta x\,\Delta p_x \geq \frac{h}{4\pi} \tag{27.16}$$

- Simultaneous measurements of **energy and lifetime** with uncertainties ΔE and Δt.

$$\Delta E\,\Delta t \geq \frac{h}{4\pi} \tag{27.17}$$

REVIEW CHECKLIST

▷ State the assumptions made by Planck in order to explain the shape of the Intensity vs Wavelength curve for a blackbody radiator.

▷ Describe the manner in which the Intensity vs Wavelength curve for a blackbody radiator changes as the temperature of the radiator increases.

▷ Describe the Einstein model for the photoelectric effect and the predictions of the fundamental photoelectric effect equation (Equation 27.6) for the maximum kinetic energy of photoelectrons. Recognize that Einstein's model of the photoelectric effect involves the photon concept $(E = hf)$, and the fact that the basic features of the photoelectric effect are consistent with this model.

▷ Describe the Compton effect (the scattering of x-rays by electrons) and be able to use the formula for the Compton shift (Eq. 27.11). Recognize that the Compton effect can only be explained using the photon concept.

▷ Use the de Broglie equation to calculate the wavelength associated with moving particles.

▷ Identify experimental evidence for the wave and particle nature of light and also for the wave nature of particles.

▷ Discuss the manner in which the uncertainty principle makes possible a better understanding of the dual wave-particle nature of light and matter.

SOLUTIONS TO SELECTED END-OF-CHAPTER PROBLEMS

1. (a) What is the surface temperature of Betelgeuse, a red giant star in the constellation of Orion, which radiates with a peak wavelength of about 970 nm? (b) Rigel, a bluish-white star in Orion, radiates with a peak wavelength of 145 nm. Find the temperature of Rigel's surface.

Solution

If we assume that the surface of a star approximates a blackbody radiator, the absolute temperature of the surface and the peak wavelength of the radiation from that star are related by Wien's displacement law,

$$\lambda_{max} T = 0.289\ 8 \times 10^{-2}\ \text{m} \cdot \text{K}$$

(a) For Betelgeuse, a red giant star, $\lambda_{max} = 970\ \text{nm}$. Thus, the surface temperature is

$$T = \frac{0.289\ 8 \times 10^{-2}\ \text{m} \cdot \text{K}}{\lambda_{max}} = \frac{0.289\ 8 \times 10^{-2}\ \text{m} \cdot \text{K}}{970 \times 10^{-9}\ \text{m}} = 2.99 \times 10^{3}\ \text{K} \approx 3\ 000\ \text{K} \qquad \Diamond$$

(b) The peak wavelength from Rigel is 145 nm, giving a surface temperature of

$$T = \frac{0.289\ 8 \times 10^{-2}\ \text{m} \cdot \text{K}}{\lambda_{max}} = \frac{0.289\ 8 \times 10^{-2}\ \text{m} \cdot \text{K}}{145 \times 10^{-9}\ \text{m}} = 2.00 \times 10^{4}\ \text{K} \approx 20\ 000\ \text{K} \qquad \Diamond$$

7.   An FM radio transmitter has a power output of 150 kW and operates at a frequency of 99.7 MHz. How many photons per second does the transmitter emit?

Solution

The power output from the transmitter is

$$\mathcal{P} = 150 \text{ kW} = 150 \times 10^3 \text{ J/s}$$

Each photon emitted by the transmitter carries away a quantity of energy given by

$$E_\gamma = hf = \left(6.63 \times 10^{-34} \text{ J·s}\right)\left(99.7 \times 10^6 \text{ Hz}\right) = 6.61 \times 10^{-26} \text{ J}$$

Thus, the number of photons emitted in an interval of $\Delta t = 1.00$ s is

$$N = \frac{\Delta E}{E_\gamma} = \frac{\mathcal{P} \cdot (\Delta t)}{E_\gamma} = \frac{\left(150 \times 10^3 \text{ J/s}\right)(1.00 \text{ s})}{6.61 \times 10^{-26} \text{ J}} = 2.27 \times 10^{30} \qquad \Diamond$$

11.   When light of wavelength 350 nm falls on a potassium surface, electrons having a maximum kinetic energy of 1.31 eV are emitted. Find (a) the work function of potassium, (b) the cutoff wavelength, and (c) the frequency corresponding to the cutoff wavelength.

Solution

(a)   Einstein's photoelectric effect equation is $KE_{max} = E_\gamma - \phi$, where KE_{max} is the maximum kinetic energy of the emitted electrons, E_γ is the energy of the photons incident on the surface, and ϕ is the work function of the material of the surface. If the wavelength of the incident light is $\lambda = 350$ nm,

$$E_\gamma = hf = \frac{hc}{\lambda} = \frac{\left(6.63 \times 10^{-34} \text{ J} \cdot \text{s}\right)\left(3.00 \times 10^8 \text{ m/s}\right)}{350 \times 10^{-9} \text{ m}} = 5.68 \times 10^{-19} \text{ J}$$

or $$E_\gamma = \left(5.68 \times 10^{-19} \text{ J}\right)\left(\frac{1.00 \text{ eV}}{1.60 \times 10^{-19} \text{ J}}\right) = 3.55 \text{ eV}$$

The observed maximum kinetic energy of the emitted electrons is $KE_{max} = 1.31$ eV. Thus, the work function of the potassium surface is

$$\phi = E_\gamma - KE_{max} = 3.55 \text{ eV} - 1.31 \text{ eV} = 2.24 \text{ eV}$$ ◊

(b)   The cutoff wavelength is the wavelength of the lowest energy photons capable of freeing electrons from the surface. At the cutoff wavelength $\left(\lambda = \lambda_c\right)$, the freed electrons leave the surface with zero kinetic energy $\left(\text{that is, } KE_{max} = 0\right)$.

From the photoelectric effect equation, $0 = \left(E_\gamma\right)_{min} - \phi$

or $\left(E_\gamma\right)_{min} = \dfrac{hc}{\lambda_c} = \phi$ and $\lambda_c = \dfrac{hc}{\phi}$

Thus, $$\lambda_c = \frac{\left(6.63 \times 10^{-34} \text{ J} \cdot \text{s}\right)\left(3.00 \times 10^8 \text{ m/s}\right)}{2.24 \text{ eV}\left(1.60 \times 10^{-19} \text{ J}/1.00 \text{ eV}\right)} = 5.55 \times 10^{-7} \text{ m} = 555 \text{ nm}$$ ◊

(c)   The frequency corresponding to the cutoff wavelength is

$$f_c = \frac{c}{\lambda_c} = \frac{3.00 \times 10^8 \text{ m/s}}{5.55 \times 10^{-7} \text{ m}} = 5.41 \times 10^{14} \text{ Hz}$$ ◊

18.   Ultraviolet light is incident normally on the surface of a certain substance. The binding energy of the electrons in this substance is 3.44 eV. The incident light has an intensity of 0.055 W/m^2. The electrons are photoelectrically emitted with a maximum speed of 4.2×10^5 m/s. How many electrons are emitted from a square centimeter of the surface each second? Assume that the absorption of every photon ejects an electron.

Solution

The total energy absorbed by a square centimeter of the surface in 1.0 s is

$$E = \mathcal{P}(\Delta t) = IA(\Delta t) = \left(0.055 \ \frac{J/s}{m^2}\right)\left[1.0 \ cm^2\left(\frac{1 \ m^2}{10^4 \ cm^2}\right)\right](1.0 \ s) = 5.5 \times 10^{-6} \ J$$

Each electron leaving the surface spends a minimum of $\phi = 3.44$ eV of energy overcoming the forces binding it to the surface. The most energetic of these electrons (the ones that spend the minimum energy to escape) leave the surface with a kinetic energy of

$$KE_{max} = \frac{1}{2}m_e v_{max}^2 = \frac{1}{2}\left(9.11 \times 10^{-31} \ kg\right)\left(4.2 \times 10^5 \ m/s\right)^2 = 8.0 \times 10^{-20} \ J$$

From Einstein's photoelectric effect equation, $KE_{max} = E_\gamma - \phi$, the energy of the absorbed photon must be

$$E_\gamma = KE_{max} + \phi = 8.0 \times 10^{-20} \ J + 3.44 \ eV\left(\frac{1.6 \times 10^{-19} \ J}{1 \ eV}\right) = 6.3 \times 10^{-19} \ J$$

Thus, the number of photons absorbed by a square centimeter of the surface each second and, assuming each absorbed photon ejects an electron, the number of electrons emitted each second is

$$N = \frac{E}{E_\gamma} = \frac{5.5 \times 10^{-6} \ J}{6.3 \times 10^{-19} \ J} = 8.7 \times 10^{12}$$

◊

23. Potassium iodide has an interplanar spacing of $d = 0.296$ nm. A monochromatic x-ray beam shows a first-order diffraction maximum when the grazing angle is 7.6°. Calculate the x-ray wavelength.

Solution

The reflection of x-rays by the regular array of atoms in a crystal produces a diffraction pattern with strong reflections (constructive interference or maxima) occurring when the incident x-rays strike the planes or layers of atoms at certain grazing angles, θ, as shown at the right. The angles at which the intensity of the reflections is a maximum are given by Bragg's law as

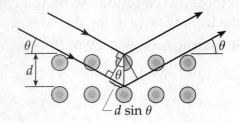

$$2d \sin \theta = m\lambda \qquad (m = 1, 2, 3, \ldots)$$

where d is the interplanar spacing, and λ is the wavelength of the x-rays.

Thus, if the first-order ($m = 1$) maximum occurs at a grazing angle of $\theta = 7.6°$ when x-rays strike a crystal having interplanar spacing $d = 0.296$ nm, the wavelength of the x-rays is

$$\lambda = \frac{2d \sin \theta}{m} = \frac{2(0.296 \text{ nm}) \sin (7.6°)}{1} = 0.078 \text{ nm} \qquad \diamond$$

29.   A 0.0016-nm photon scatters from a free electron. For what (photon) scattering angle will the recoiling electron and scattered photon have the same kinetic energy?

Solution

Before the scattering event, the energy of the free electron was just its rest energy E_R. After the event, the total energy of the electron is

$$E = KE + E_R$$

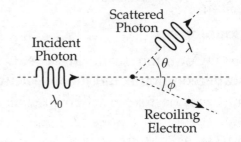

Scattered Photon

Incident Photon

Recoiling Electron

The energy of the incident photon is $E_\gamma = hc/\lambda_0$, while that of the scattered photon is $E_\gamma' = hc/\lambda$. Conservation of energy then gives

$$E_\gamma + E_R = E_\gamma' + KE + E_R \qquad \text{or} \qquad E_\gamma = E_\gamma' + KE$$

Thus, if the kinetic energy of the recoiling electron is the same as the energy of the scattered photon, we have $E_\gamma = E_\gamma' + E_\gamma' = 2E_\gamma'$ giving

$$\frac{hc}{\lambda_0} = 2\left(\frac{hc}{\lambda}\right) \qquad \text{or} \qquad \lambda = 2\lambda_0$$

The Compton shift formula, $\Delta\lambda = \lambda - \lambda_0 = \dfrac{h}{m_e c}(1-\cos\theta)$, then gives the photon scattering angle as

$$\cos\theta = 1 - \left(\frac{m_e c}{h}\right)\Delta\lambda = 1 - \left(\frac{m_e c}{h}\right)(2\lambda_0 - \lambda_0) = 1 - \left(\frac{m_e c}{h}\right)\lambda_0$$

or $\quad \theta = \cos^{-1}\left[1 - \dfrac{(9.11\times10^{-31}\ \text{kg})(3.00\times10^8\ \text{m/s})}{6.63\times10^{-34}\ \text{J}\cdot\text{s}}(0.0016\times10^{-9}\text{m})\right] = 70°$ ⬦

31. A 0.110-nm photon collides with a stationary electron. After the collision, the electron moves forward, and the photon recoils backward. Find the momentum and kinetic energy of the electron.

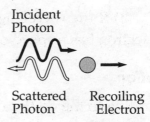

Incident Photon

Scattered Photon   Recoiling Electron

Solution

In a Compton scattering event, the difference between the wavelength λ of the scattered photon and the wavelength λ_0 of the incident photon is given by

$$\Delta\lambda = \lambda - \lambda_0 = \lambda_C (1 - \cos\theta)$$

where $\lambda_C = h/m_e c = 0.002\,43$ nm is the Compton wavelength, and θ is the scattering angle.

When $\theta = 180°$ (that is, the scattered photon moves in the backward direction), this gives

$$\Delta\lambda = \lambda_C (1 - \cos 180°) = 2\lambda_C = 0.004\,86 \text{ nm}$$

If $\lambda_0 = 0.110$ nm, the wavelength of the scattered photon is

$$\lambda = \lambda_0 + \Delta\lambda = 0.110 \text{ nm} + 0.004\,86 \text{ nm} = 0.115 \text{ nm}$$

The kinetic energy of the recoiling electron is then

$$KE = \left(E_\gamma\right)_0 - E_\gamma = \frac{hc}{\lambda_0} - \frac{hc}{\lambda} = \frac{hc(\Delta\lambda)}{\lambda_0\lambda}$$

$$= \frac{\left(6.63\times10^{-34} \text{ J·s}\right)\left(3.00\times10^8 \text{ m/s}\right)\left(0.004\,86 \text{ nm}\right)}{\left(0.110\times10^{-9} \text{ m}\right)\left(0.115 \text{ nm}\right)} = 7.65\times10^{-17} \text{ J}$$

$$KE = \left(7.65\times10^{-17} \text{ J}\right)\left(1 \text{ eV}/1.60\times10^{-19} \text{ J}\right) = 478 \text{ eV} \qquad \Diamond$$

Since this electron is non-relativistic $\left(KE \ll E_R = 0.511 \text{ MeV}\right)$, its momentum is given by $p = m_e v = \sqrt{2m_e\left(\tfrac{1}{2}m_e v^2\right)} = \sqrt{2m_e(KE)}$

$$p = \sqrt{2\left(9.11\times10^{-31} \text{ kg}\right)\left(7.65\times10^{-17} \text{ J}\right)} = 1.18\times10^{-23} \text{ kg·m/s} \qquad \Diamond$$

35.   (a) If the wavelength of an electron is 5.00×10^{-7} m, how fast is it moving? (b) If the electron has a speed of 1.00×10^{7} m/s, what is its wavelength?

Solution

According to the de Broglie hypothesis, material particles of momentum p should also have wave properties with a wavelength given by $\lambda = h/p$. For non-relativistic particles, $p = mv$ and the de Broglie wavelength is $\lambda = h/mv$.

(a)   If an electron has a wavelength of $\lambda = 5.00 \times 10^{-7}$ m, its speed will be

$$v = \frac{h}{m_e \lambda} = \frac{6.63 \times 10^{-34}\ \text{J} \cdot \text{s}}{\left(9.11 \times 10^{-31}\ \text{kg}\right)\left(5.00 \times 10^{-7}\ \text{m}\right)} = 1.46 \times 10^{3}\ \text{m/s} = 1.46\ \text{km/s} \qquad \Diamond$$

(b)   If the electron has a speed of $v = 1.00 \times 10^{7}$ m/s $= 0.033\ 3c$, its wavelength is

$$\lambda = \frac{h}{m_e v} = \frac{6.63 \times 10^{-34}\ \text{J} \cdot \text{s}}{\left(9.11 \times 10^{-31}\ \text{kg}\right)\left(1.00 \times 10^{7}\ \text{m/s}\right)} = 7.28 \times 10^{-11}\ \text{m} \qquad \Diamond$$

41. The resolving power of a microscope is proportional to the wavelength used. A resolution of 1.0×10^{-11} m (0.010 nm) would be required in order to "see" an atom. (a) If electrons were used (electron microscope), what minimum kinetic energy would be required of the electrons? (b) If photons were used, what minimum photon energy would be needed to obtain 1.0×10^{-11} m resolution?

Solution

(a)   The de Broglie wavelength of an electron is $\lambda = h/p$ where p is the linear momentum of the electron. If this wavelength is to be $\lambda = 1.0\times10^{-11}$ m, the electron must have a momentum of

$$p = \frac{h}{\lambda} = \frac{6.63\times10^{-34}\ \text{J}\cdot\text{s}}{1.0\times10^{-11}\ \text{m}} = 6.6\times10^{-23}\ \text{kg}\cdot\text{m/s}$$

Allowing for the possibility that the electron (with rest energy $E_R = 511$ keV) might be relativistic, we compute its total energy as

$$E = \sqrt{(pc)^2 + E_R^2} = \sqrt{\left[\frac{(6.6\times10^{-23}\ \text{kg}\cdot\text{m/s})(3.00\times10^8\ \text{m/s})}{1.60\times10^{-16}\ \text{J/keV}}\right]^2 + (511\ \text{keV})^2}$$

This yields $E = 526$ keV, so the kinetic energy of the electron is

$$KE = E - E_R = 526\ \text{keV} - 511\ \text{keV} = 15\ \text{kev} \qquad \Diamond$$

Thus, we see that the electron is not relativistic, and classical methods could have been used to obtain

$$KE = \frac{p^2}{2m_e} = \frac{(6.6\times10^{-23}\ \text{kg}\cdot\text{m/s})^2}{2(9.11\times10^{-31}\ \text{kg})}\left(\frac{1\ \text{keV}}{1.60\times10^{-16}\ \text{J}}\right) = 15\ \text{keV}$$

(b)   The energy of a photon having a wavelength of $\lambda = 1.0\times10^{-11}$ m is

$$E_\gamma = hf = \frac{hc}{\lambda}$$

$$= \frac{(6.63\times10^{-34}\ \text{J}\cdot\text{s})(3.00\times10^8\ \text{m/s})}{1.0\times10^{-11}\ \text{m}}\left(\frac{1\ \text{keV}}{1.60\times10^{-16}\ \text{J}}\right) = 1.2\times10^2\ \text{keV} \qquad \Diamond$$

45.   Suppose optical radiation $\left(\lambda = 5.00 \times 10^{-7}\text{ m}\right)$ is used to determine the position of an electron to within the wavelength of the light. What will be the resulting uncertainty in the electron's velocity?

Solution

If the position is determined to within the wavelength of the light, the uncertainty in the position of the electron is $\Delta x = \lambda$. The uncertainty principle, $(\Delta x)(\Delta p_x) \geq h/4\pi$, places a lower limit on the uncertainty in the linear momentum as

$$\Delta p_x \geq \frac{h}{4\pi (\Delta x)} = \frac{h}{4\pi\lambda}$$

Assuming that the electron is non-relativistic, its linear momentum is $p_x = m_e v$. Thus, the uncertainty in momentum is $\Delta p_x = \Delta(m_e v) = m_e (\Delta v)$, and the uncertainty in the electron's velocity is

$$\Delta v = \frac{\Delta p_x}{m_e} = \frac{h/4\pi\lambda}{m_e} = \frac{h}{4\pi m_e \lambda}$$

With $\lambda = 5.00 \times 10^{-7}$ m and $m_e = 9.11 \times 10^{-31}$ kg,

this gives $\Delta v = \dfrac{h}{4\pi m_e \lambda} = \dfrac{6.63 \times 10^{-34}\text{ J}\cdot\text{s}}{4\pi\left(9.11 \times 10^{-31}\text{ kg}\right)\left(5.00 \times 10^{-7}\text{ m}\right)} = 116\text{ m/s}$ ◊

49.   The spacing between planes of nickel atoms in a nickel crystal is 0.352 nm. At what angle does a second-order Bragg reflection occur in nickel for 11.3-keV x-rays?

Solution

When x-rays reflect from the regularly spaced layers of atoms within a crystal, the angles (measured from the planes of the atom layers) where constructive interference, or maxima in the reflected radiation, occurs are given by Bragg's law as

$$2d\sin\theta = m\lambda \qquad m = 1,2,3,\ldots$$

Here, d is the spacing between the planes of the atom layers, and λ is the wavelength of the radiation. The wavelength of 11.3-keV x-rays is

$$\lambda = \frac{c}{f} = \frac{c}{E_\gamma/h} = \frac{hc}{E_\gamma} = \frac{\left(6.63\times10^{-34}\ \text{J}\cdot\text{s}\right)\left(3.00\times10^{8}\ \text{m/s}\right)}{11.3\ \text{keV}\left(1.60\times10^{-16}\ \text{J/keV}\right)} = 1.10\times10^{-10}\ \text{m}$$

Thus, if $d = 0.352$ m, the angle where the second-order $(m = 2)$ Bragg reflection occurs is

$$\theta = \sin^{-1}\left(\frac{m\lambda}{2d}\right) = \sin^{-1}\left[\frac{2\left(1.10\times10^{-10}\ \text{m}\right)}{2\left(0.352\times10^{-9}\ \text{m}\right)}\right] = 18.2° \qquad \Diamond$$

51.   Photons of wavelength 450 nm are incident on a metal. The most energetic electrons ejected from the metal are bent into a circular arc of radius 20.0 cm by a magnetic field with a magnitude of 2.00×10^{-5} T. What is the work function of the metal?

Solution

When electrons move perpendicularly to a magnetic field, they are deflected into a circular path with the required centripetal acceleration produced by the magnetic force acting on them. Thus,

$$m_e \left(\frac{v^2}{r} \right) = evB \qquad \text{or} \qquad v = \frac{eBr}{m_e}$$

The kinetic energy of the most energetic electrons ejected from the metal is

$$KE_{max} = \frac{1}{2} m_e v^2 = \frac{e^2 B^2 r^2}{2m_e} = \frac{\left(1.60 \times 10^{-19}\ \text{C}\right)^2 \left(2.00 \times 10^{-5}\ \text{T}\right)^2 \left(0.200\ \text{m}\right)^2}{2\left(9.11 \times 10^{-31}\ \text{kg}\right)}$$

or $KE_{max} = 2.25 \times 10^{-19}\ \text{J} \left(\dfrac{1\ \text{eV}}{1.60 \times 10^{-19}\ \text{J}} \right) = 1.40\ \text{eV}$

The energy of the photons incident on the metal surface is

$$E_\gamma = hf = \frac{hc}{\lambda} = \frac{\left(6.63 \times 10^{-34}\ \text{J} \cdot \text{s}\right)\left(3.00 \times 10^8\ \text{m/s}\right)}{450 \times 10^{-9}\ \text{m}} \left(\frac{1\ \text{eV}}{1.60 \times 10^{-19}\ \text{J}} \right) = 2.76\ \text{eV}$$

Einstein's photoelectric effect equation, $KE_{max} = E_\gamma - \phi$, then gives the work function of the metal as

$$\phi = E_\gamma - KE_{max} = 2.76\ \text{eV} - 1.40\ \text{eV} = 1.36\ \text{eV}$$ ◊

55. How fast must an electron be moving if all its kinetic energy is lost to a single x-ray photon (a) at the high end of the x-ray electromagnetic spectrum with a wavelength of 1.00×10^{-8} m; (b) at the low end of the x-ray electromagnetic spectrum with a wavelength of 1.00×10^{-13} m?

Solution

Photons with a wavelength of $\lambda = 1.00 \times 10^{-8}$ m have energy

$$E_\gamma = hf = \frac{hc}{\lambda} = \frac{(6.63 \times 10^{-34} \text{ J} \cdot \text{s})(3.00 \times 10^8 \text{ m/s})}{1.00 \times 10^{-8} \text{ m}} \left(\frac{1 \text{ eV}}{1.60 \times 10^{-19} \text{ J}} \right) = 124 \text{ eV}$$

and those with $\lambda' = 10^{-5}\lambda = 1.00 \times 10^{-13}$ m have energy $E'_\gamma = 10^5 E_\gamma = 12.4$ MeV

(a) If an electron (with rest energy $E_R = 0.511$ MeV) has kinetic energy $KE = E_\gamma = 124$ eV, then $KE \ll E_R$ and the electron is non-relativistic. The speed of the electron is then

$$v = \sqrt{\frac{2\,KE}{m_e}} = \sqrt{\frac{2(124 \text{ eV})}{9.11 \times 10^{-31} \text{ kg}} \left(\frac{1.60 \times 10^{-19} \text{ J}}{1 \text{ eV}} \right)} = 6.61 \times 10^6 \text{ m/s} = 0.022\,0c \qquad \lozenge$$

(b) An electron having $KE = E'_\gamma = 12.4$ MeV is highly relativistic, and its total energy is $E = \gamma E_R = KE + E_R$, which gives

$$\gamma = 1 + \frac{KE}{E_R} = 1 + \frac{12.4 \text{ MeV}}{0.511 \text{ MeV}} = 25.3$$

Since $\gamma = \dfrac{1}{\sqrt{1 - (v/c)^2}}$, the speed of the electron must be

$$v = c\sqrt{1 - \frac{1}{\gamma^2}} = c\sqrt{1 - \frac{1}{(25.3)^2}} = 0.999\,2\,c \qquad \lozenge$$

Chapter 28
Atomic Physics

NOTES FROM SELECTED CHAPTER SECTIONS

28.2   Atomic Spectra

28.3   The Bohr Theory of Hydrogen

The basic postulates of the Bohr model of the hydrogen atom are:

- The **electron moves in circular orbits** about the nucleus under the influence of the Coulomb force of attraction between the electron and the positively charged proton in the nucleus.

- The **electron can exist only in very specific stationary states** (orbits). When the electron is in one of its allowed orbits, it does not emit energy by radiation.

- The **atom radiates energy only when the electron makes a transition** ("jumps") from one allowed stationary orbit to another less energetic state. This postulate states that the energy given off by an atom is carried away by a photon of energy hf.

- The **orbital angular momentum of the electron about the nucleus is quantized** (an integer multiple of $\hbar = \dfrac{h}{2\pi}$, where h is Planck's constant). *This condition determines the radii of the allowed orbits.*

There are several important reasons to understand the behavior of the hydrogen atom as an atomic system:

- Much of what is learned about the hydrogen atom with its single electron can be extended to such single-electron ions as He^+ and Li^{2+}, which are hydrogen-like in their atomic structure.

- The hydrogen atom is an ideal system for performing precise tests of theory against experiment and for improving our overall understanding of atomic structure.

- The quantum numbers used to characterize the allowed states of hydrogen can be used to describe the allowed states of more complex atoms. This enables us to understand the periodic table of the elements.

- The basic ideas about atomic structure must be well understood before we attempt to deal with the complexities of molecular structures and the electronic structure of solids.

28.4   Modification of the Bohr Theory
28.6   Quantum Mechanics and the Hydrogen Atom
28.7   The Spin Magnetic Quantum Number

The possible stationary energy states of an electron in an atom are determined by the values of four quantum numbers.

- **Principle quantum number, n; integer values from 1 to ∞**

The principle quantum number follows from the concept of quantization of angular momentum. **Electron energy states with the same principle quantum number form a shell** identified by the letters $K, L, M \ldots$ corresponding to $n = 1, 2, 3 \ldots$

- **Orbital quantum number, ℓ; integer values from 0 to $(n-1)$**

An electron in a given allowed energy state may move in different elliptical orbits determined by the value of ℓ. For each value of n there are n possible orbits corresponding to different values of ℓ. **Energy states with given values of n and ℓ form a subshell** identified by the letters $s, p, d, f, \ldots$ corresponding to $\ell = 0, 1, 2, 3, \ldots$. The maximum number of electrons allowed in any subshell is $2(2\ell + 1)$.

- **Orbital magnetic quantum number, m_ℓ; integer values from $-\ell$ to $+\ell$**

The orbital magnetic quantum number accounts for the observed **Zeeman effect**; when a gas is placed in an external magnetic field, single spectral lines are split into several lines.

- **Spin magnetic quantum number, m_s ; values of –1/2 and +1/2**

  The **spin magnetic quantum number** m_s accounts for the two closely spaced energy states in spectral lines ("doublets") corresponding to the two possible orientations of electron spin (+1/2 is "up" spin and –1/2 is "down" spin).

28.9 The Exclusion Principle and the Periodic Table

At this point you might find it useful to review the description of the quantum numbers in Section 28.4 of the textbook.

The Pauli exclusion principle states that no two electrons in an atom can exist in identical quantum states. This means that no two electrons in a given atom can have exactly the same values for the set of quantum numbers n, ℓ, m_ℓ and m_s.

The order in which electrons fill subshells and the application of the exclusion principle are illustrated in the following tables in your textbook:

Table 28.3, Number of Electrons in Filled Subshells and Shells

Table 28.4, Electronic Configuration of Some Elements

28.10 Characteristic x-Rays

The x-ray spectrum of a metal target consists of a broad continuous spectrum on which are superimposed a series of sharp lines. **Characteristic x-rays** are emitted by atoms when an electron undergoes a transition from an outer shell into an electron vacancy in one of the inner shells. Transitions into a vacant state in the K shell give rise to the K series of spectral lines, transitions into a vacant state in the L shell create the L series of lines, and so on. **Lines within a series are given a notation to designate the shell from which the transition originates. Examples from the K and L series (in order of increasing energy and decreasing intensity) are:**

- K_α line; an electron drops from the L shell to the K shell

- K_β line; an electron drops from the M shell to the K shell

- L_α line; an electron drops from the M shell to the L shell

- L_β line; an electron drops from the N shell to the L shell

28.11 Atomic Transitions
28.12 Lasers and Holography

Stimulated absorption (upward atomic transition) occurs when photons incident on a gas have energies which exactly match the energy separation between two allowed states of an atom, usually the ground state and a higher energy state. The absorption process raises an atom to various higher energy states called **excited states**.

An atom in an excited state has a certain probability of returning to its original energy state by emission of a photon. This process is called **spontaneous emission.**

When a photon with an energy equal to the excitation energy of an excited atom is incident on the atom, it can increase the probability of de-excitation. This is called **stimulated emission (downward atomic transition)** and results in a second photon. The incident and the emitted photons are identical; they have equal energies, and they are exactly in phase.

Population inversion in a gas sample is the condition in which there are more atoms in the excited state than there are in the ground state.

Lasers are monochromatic, coherent light sources that work on the principle of stimulated emission of radiation by a system of atoms.

The following conditions must be satisfied in order to achieve laser action:

- **Population inversion.** There must be a greater number of atoms in an excited state than in the ground state.

- **Metastable states.** The lifetime of the excited states must be long compared with the usually short lifetimes of excited states. Under these conditions, stimulated emission will occur before spontaneous emission.

- **Photon confinement.** The emitted photons must remain in the system long enough to allow them to stimulate further emission from other excited atoms. This is achieved by the use of reflecting mirrors at the ends of the system. One end is made totally reflecting, and the other is slightly transparent to allow the laser beam to escape.

EQUATIONS AND CONCEPTS

Wavelengths in the Balmer series in the emission spectrum of hydrogen can be calculated from an empirical equation.

$$\frac{1}{\lambda} = R_H\left(\frac{1}{2^2} - \frac{1}{n^2}\right)$$
$$(n = 3, 4, 5, \ldots)$$

(28.1)

R_H is a constant called the **Rydberg constant.**

$$R_H = 1.097\,373\,2 \times 10^7 \text{ m}^{-1}$$

(28.2)

An **electron transition** from an initial stationary state, E_i, to a lower energy state, E_f, results in the emission of a photon with a frequency that is proportional to the difference in energies of the initial and final states. While in one of the allowed orbits (stationary states) determined by quantization of the orbital angular momentum, the electron does not radiate energy.

$$E_i - E_f = hf$$

(28.3)

The angular momentum of the electron about the nucleus must be quantized, always equal to an integral multiple of $\hbar$. *This condition determines the radii of the allowed electron orbits.*

$$m_e vr = n\hbar$$
$$(n = 1, 2, 3, \ldots)$$

(28.4)

The total energy of the hydrogen atom (*KE* plus *PE* of the proton-electron bound system) depends on the radius of the allowed orbit of the electron. Note that the total energy of the atom is negative for all values of r except $r = \infty$ when $E = 0$.

$$E = -\frac{k_e e^2}{2r}$$

(28.8)

The electron can exist only in certain allowed orbits. When $n = 1$, r is equal to the **Bohr radius**, a_0. The radii of all allowed orbits in hydrogen can be stated in terms of a_0.

$$r_n = \frac{n^2 \hbar^2}{m_e k_e e^2} \qquad (28.9)$$

$$(n = 1, 2, 3, \ldots)$$

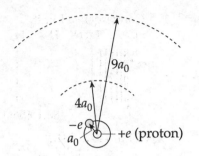

$$a_0 = \frac{\hbar^2}{m_e k_e e^2} = 0.052\ 9 \text{ nm} \qquad (28.10)$$

$$r_n = n^2 a_0 = n^2 (0.052\ 9 \text{ nm}) \qquad (28.11)$$

The energies of the quantum states of hydrogen are given by Equation 28.12.

$$E_n = -\frac{m_e k_e^2 e^4}{2\hbar^2} \left(\frac{1}{n^2}\right) \qquad (28.12)$$

Quantized energy level values can be expressed in units of electron volts (eV). *The lowest allowed energy state or ground state corresponds to the principal quantum number $n=1$. The absolute value of the ground state energy is equal to the ionization energy of the atom. The energy level approaches $E=0$ as r approaches infinity.*

$$E_n = -\frac{13.6}{n^2} \text{ eV} \qquad (28.13)$$

A **photon** of frequency f (and wavelength λ) is emitted when an electron undergoes a transition from an initial energy level E_i to a lower level E_f.

$$f = \frac{E_i - E_f}{h} = \frac{m_e k_e^2 e^4}{4\pi \hbar^3} \left(\frac{1}{n_f^2} - \frac{1}{n_i^2}\right) \qquad (28.14)$$

$$\text{where } n_f < n_i$$

A **theoretical value for the Rydberg constant** can be calculated by substituting known values into Equation 28.16. The value of R obtained from Equation 28.16 is in excellent agreement with the experimental value given in Equation 28.2.

$$R = \frac{m_e k_e^2 e^4}{4\pi c \hbar^3} \qquad (28.16)$$

Photon wavelengths can be calculated using Equation 28.17. Wavelength, rather than frequency, is the experimentally measured quantity.

$$\frac{1}{\lambda} = R\left(\frac{1}{n_f^2} - \frac{1}{n_i^2}\right) \qquad (28.17)$$

Prominent lines in the visible region of the Balmer series of hydrogen are shown in the figure.

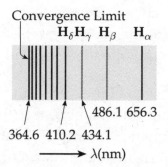

The **observed spectral lines** in the hydrogen spectrum can be arranged into several series. Each series corresponds to an assigned value of the principal quantum number of the final energy state (n_f).

Lyman series: $n_f = 1;\ \ n_i = 2, 3, 4, \ldots$
Balmer series: $n_f = 2;\ \ n_i = 3, 4, 5, \ldots$
Paschen series: $n_f = 3;\ \ n_i = 4, 5, 6, \ldots$
Brackett series: $n_f = 4;\ \ n_i = 5, 6, 7, \ldots$

Quantum mechanics is in agreement with classical physics when the energy differences between quantized levels are small. For very large values of the principal quantum number n the energy differences between adjacent levels approach zero and essentially a continuous range (as opposed to a quantized set) of energy values of the emitted photons is observed.

The correspondence principle

The **current-voltage relationship for an ideal diode** depends on the direction of the applied voltage. For forward bias (when the p side of the junction is positive) the current increases exponentially with voltage; a reverse bias (the n side of the junction is positive) results in a small reverse current that reaches saturation value I_0. In the exponent of Equation 28.21, q is the electron charge.

$$I = I_0(e^{q\Delta V/k_B T} - 1) \qquad (28.21)$$

T = Kelvin temperature

REVIEW CHECKLIST

▷ State the basic postulates of the Bohr model of the hydrogen atom.

▷ Sketch an energy level diagram for hydrogen (assign values of the principle quantum number, n), show transitions corresponding to spectral lines in the several known series, and make calculations of wavelength values.

▷ For each of the quantum numbers, n (the principal quantum number), ℓ (the orbital quantum number), m_ℓ (the orbital magnetic quantum number), and m_s (the spin magnetic quantum number): (i) qualitatively describe what each implies concerning atomic structure, (ii) state the allowed values which may be assigned to each, and (iii) give the number of allowed states which may exist in a particular atom corresponding to each allowed value of the principle quantum number.

▷ State the Pauli exclusion principle and describe its relevance to the periodic table of the elements. Show how the exclusion principle leads to the known electronic ground state configuration of the light elements.

SOLUTIONS TO SELECTED END-OF-CHAPTER PROBLEMS

5. The "size" of the atom in Rutherford's model is about 1.0×10^{-10} m. (a) Determine the speed of an electron moving about the proton using the attractive electrical force between an electron and a proton separated by this distance. (b) Does this speed suggest that Einsteinian relativity must be considered in studying the atom? (c) Compute the de Broglie wavelength of the electron as it moves about the proton. (d) Does this wavelength suggest that wave effects, such as diffraction and interference, must be considered in studying the atom?

Solution

(a)   The force of electrical attraction between the electron and proton has a magnitude $F = k_e e^2 / r^2$ and produces the needed centripetal acceleration, $a_c = v^2 / r$, as the electron orbits the proton. Thus, we can find the orbital speed of the electron by applying Newton's second law to the orbital motion:

$$m_e \frac{v^2}{r} = \frac{k_e e^2}{r^2}$$

or, isolating the velocity,

$$v = \sqrt{\frac{k_e e^2}{m_e r}}$$

If the electron and proton are separated by $r = 1.0 \times 10^{-10}$ m

the speed is $v = \sqrt{\dfrac{\left(8.99 \times 10^9 \text{ N} \cdot \text{m}^2 / \text{C}^2\right)\left(1.60 \times 10^{-19} \text{ C}\right)^2}{\left(9.11 \times 10^{-31} \text{ kg}\right)\left(1.0 \times 10^{-10} \text{ m}\right)}} = 1.6 \times 10^6$ m/s   ◊

(b)   Since $v/c = \left(1.6 \times 10^6 \text{ m/s}\right) / \left(3.0 \times 10^8 \text{ m/s}\right) \ll 1$

the electron is not relativistic, and relativity effects need not be considered.   ◊

(c)   The de Broglie wavelength of the electron is $\lambda = \dfrac{h}{p} = \dfrac{h}{m_e v}$

Therefore, $\lambda = \dfrac{6.63 \times 10^{-34} \text{ J} \cdot \text{s}}{\left(9.11 \times 10^{-31} \text{ kg}\right)\left(1.6 \times 10^6 \text{ m/s}\right)} = 4.6 \times 10^{-10}$ m   ◊

(d)   Since the de Broglie wavelength of the electron and the size of the atom are of the same order of magnitude, wave effects such as interference and diffraction should be considered when studying the atom.   ◊

7.   A hydrogen atom is in its first excited state $(n = 2)$. Using the Bohr theory of the atom, calculate (a) the radius of the orbit, (b) the linear momentum of the electron, (c) the angular momentum of the electron, (d) the kinetic energy, (e) the potential energy, and (f) the total energy.

Solution

(a)   In the Bohr theory, the radii of the allowed orbits are $r_n = n^2 a_0$ where $a_0 = 0.052\,9$ nm. Thus, in the $n = 2$ state, the radius of the orbit is

$$r_2 = (2)^2 (0.052\,9 \text{ nm}) = 0.212 \text{ nm}$$ ◊

(b)   The electrical attraction between the electron and proton supplies the centripetal acceleration of the electron. Therefore,

$$m_e \frac{v_n^2}{r_n} = \frac{k_e e^2}{r_n^2} \quad \text{which yields} \quad v_n = \sqrt{\frac{k_e e^2}{m_e r_n}} \quad \text{and} \quad p_n = m_e v_n = \sqrt{\frac{m_e k_e e^2}{r_n}}$$

The linear momentum of the electron in the $n = 2$ state is then

$$p_2 = \sqrt{\frac{m_e k_e e^2}{r_2}} = \sqrt{\frac{(9.11 \times 10^{-31} \text{ kg})(8.99 \times 10^9 \text{ N} \cdot \text{m}^2/\text{C}^2)(1.60 \times 10^{-19} \text{ C})^2}{0.212 \times 10^{-9} \text{ m}}}$$

or $$p_2 = 9.95 \times 10^{-25} \text{ kg} \cdot \text{m/s}$$ ◊

(c)   According to Bohr's postulates, the angular momentum of the electron is quantized, having values given by $L_n = n\hbar = n(h/2\pi)$. Thus, in the $n = 2$ state, we have

$$L_2 = 2\left(\frac{h}{2\pi}\right) = 2\left(\frac{6.63 \times 10^{-34} \text{ J} \cdot \text{s}}{2\pi}\right) = 2.11 \times 10^{-34} \text{ J} \cdot \text{s}$$ ◊

(d)   The kinetic energy of the orbiting electron is $KE_n = \frac{1}{2} m_e v_n^2 = \frac{p_n^2}{2m_e}$

When $n = 2$, we use the result from Part (b) and find

$$KE_2 = \frac{1}{2} m_e v_2^2 = \frac{p_2^2}{2m_e} = \frac{(9.95 \times 10^{-25} \text{ kg} \cdot \text{m/s})^2}{2(9.11 \times 10^{-31} \text{ kg})}$$

or $$KE_2 = 5.44 \times 10^{-19} \text{ J}\left(\frac{1 \text{ eV}}{1.60 \times 10^{-19} \text{ J}}\right) = 3.40 \text{ eV}$$ ◊

(e) The electrical potential energy between two point charges separated by distance r is $PE = k_e q_1 q_2 / r$. Thus, for a hydrogen atom in the $n = 2$ state,

$$PE_2 = \frac{k_e(-e)(+e)}{r_2} = -\frac{(8.99 \times 10^9 \ \text{N} \cdot \text{m}^2/\text{C}^2)(1.60 \times 10^{-19} \ \text{C})^2}{0.212 \times 10^{-9} \ \text{m}}$$

or $\qquad PE_2 = -1.09 \times 10^{-18} \ \text{J}\left(\dfrac{1 \ \text{eV}}{1.60 \times 10^{-19} \ \text{J}}\right) = -6.80 \ \text{eV}$ ◊

(f) The total energy of the hydrogen when the electron is in the $n = 2$ state is

$$E_2 = KE_2 + PE_2 = +3.40 \ \text{eV} - 6.80 \ \text{eV} = -3.40 \ \text{eV}$$ ◊

14.   A hydrogen atom initially in its ground state $(n=1)$ absorbs a photon and ends up in the state for which $n=3$. (a) What is the energy of the absorbed photon? (b) If the atom eventually returns to the ground state, what photon energies could the atom emit?

Solution

(a)   The energy of the hydrogen atom when the electron is in the quantum state associated with the quantum number n is

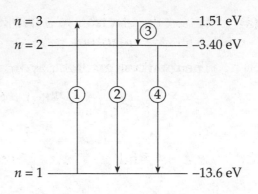

$$E_n = -\frac{13.6 \text{ eV}}{n^2} \quad \text{where} \quad n = 1, 2, 3, \ldots$$

Thus, the energies of the three lowest energy states are $E_1 = -13.6$ eV, $E_2 = -3.40$ eV, and $E_3 = -1.51$ eV as shown in the energy level diagram at the right.

If the electron is to be excited from the ground state $(n=1)$ to the second excited state $(n=3)$ by absorption of a photon as shown by transition 1 in the above diagram, the energy of the absorbed photon must be

$$E_\gamma = E_3 - E_1 = -1.51 \text{ eV} - (-13.6 \text{ eV}) = 12.1 \text{ eV} \qquad \Diamond$$

(b)   As the atom returns to the ground state after the electron is excited to the $n=3$ state, there are three possible transitions the electron could make as shown by the vertical arrows labeled 2, 3, and 4 in the above diagram. The photon energies produced by each of these transitions are

Transition 2: $E_\gamma = E_3 - E_1 = -1.51 \text{ eV} - (-13.6 \text{ eV}) = 12.1 \text{ eV}$ $\qquad \Diamond$

Transition 3: $E_\gamma = E_3 - E_2 = -1.51 \text{ eV} - (-3.40 \text{ eV}) = 1.89 \text{ eV}$ $\qquad \Diamond$

Transition 4: $E_\gamma = E_2 - E_1 = -3.40 \text{ eV} - (-13.6 \text{ eV}) = 10.2 \text{ eV}$ $\qquad \Diamond$

19. (a) If an electron makes a transition from the $n = 4$ Bohr orbit to the $n = 2$ orbit, determine the wavelength of the photon created in the process. (b) Assuming that the atom was initially at rest, determine the recoil speed of the hydrogen atom when this photon is emitted.

Solution

(a) When the electron in a hydrogen atom makes a transition from an orbit with principle quantum number n_i to an orbit with quantum number n_f, the energy of the emitted photon is

$$E_\gamma = E_{n_i} - E_{n_f} = -\frac{13.6 \text{ eV}}{n_i^2} - \left(-\frac{13.6 \text{ eV}}{n_f^2}\right) = (13.6 \text{ eV})\left(\frac{1}{n_f^2} - \frac{1}{n_i^2}\right)$$

If $n_i = 4$ and $n_f = 2$, the photon energy is

$$E_\gamma = (13.6 \text{ eV})\left(\frac{1}{2^2} - \frac{1}{4^2}\right) = 2.55 \text{ eV}$$

The wavelength of this photon is

$$\lambda = \frac{c}{f} = \frac{hc}{E_\gamma} = \frac{(6.63 \times 10^{-34} \text{ J·s})(3.00 \times 10^8 \text{ m/s})}{2.55 \text{ eV}(1.60 \times 10^{-19} \text{ J/eV})} = 4.88 \times 10^{-7} \text{ m} = 488 \text{ nm} \qquad \Diamond$$

(b) The linear momentum of the system was zero before photon emission and must therefore be zero after this event. Thus, the linear momentum of the recoiling atom and that of the photon must have equal magnitudes but opposite directions.

Equating magnitudes: $\qquad\qquad\qquad p_{atom} = p_\gamma$

or $\qquad\qquad\qquad\qquad\qquad\qquad m_{atom}v = h/\lambda$

Since, for a hydrogen atom, $\qquad\qquad m_{atom} \approx m_{proton} = 1.67 \times 10^{-27} \text{ kg}$

this gives the recoil speed of the atom as

$$v = \frac{h}{m_{atom}\lambda} = \frac{6.63 \times 10^{-34} \text{ J·s}}{(1.67 \times 10^{-27} \text{ kg})(4.88 \times 10^{-7} \text{ m})} = 0.814 \text{ m/s} \qquad \Diamond$$

25. Two hydrogen atoms, both initially in the ground state, undergo a head-on collision. If both atoms are to be excited to the $n=2$ level in this collision, what is the minimum speed each atom can have before the collision?

Solution

The energy required to excite the electrons in the atoms from the ground state $(n=1)$ to the first excited state $(n=2)$ will come from the initial kinetic energies of the atoms. The total energy required to excite both atoms to this state is

$$E_{ex} = 2(E_2 - E_1) = 2\left[-\frac{13.6 \text{ eV}}{2^2} - \left(-\frac{13.6 \text{ eV}}{1^2}\right)\right] = 20.4 \text{ eV}$$

If the atomic speeds, and hence the kinetic energies before the collision, are to have the minimum values for which this process will be possible, there will be no residual kinetic energy after the collision. Hence, $KE_f = 0$ giving $KE_i = E_{ex} = 20.4 \text{ eV}$. This also means that there will be zero total momentum after the collision, so the atoms must have equal magnitude and oppositely directed momenta before impact, meaning that

$$KE_i = 2\left(\frac{1}{2}m_{atom}v^2\right) = m_{atom}v^2$$

With $m_{atom} \approx m_{proton} = 1.67 \times 10^{-27}$ kg, this gives the required minimum initial speed of each atom as

$$v = \sqrt{\frac{KE_i}{m_{proton}}} = \sqrt{\frac{20.4 \text{ eV}}{1.67 \times 10^{-27} \text{ kg}}\left(\frac{1.60 \times 10^{-19} \text{ J}}{1 \text{ eV}}\right)} = 4.42 \times 10^4 \text{ m/s} \qquad \Diamond$$

31. Determine the wavelength of an electron in the third excited orbit of the hydrogen atom, with $n = 4$.

Solution

For a non-relativistic electron, the magnitude of the linear momentum is $p = m_e v$, and the de Broglie wavelength is

$$\lambda = \frac{h}{p} = \frac{h}{m_e v}$$

From the Bohr model, an electron in one of the allowed orbits has an angular momentum of

$$L_n = m_e v_n r_n = n\hbar = n h/2\pi$$

Thus, the magnitude of the linear momentum is $\quad m_e v_n = nh/2\pi \, r_n$

and the de Broglie wavelength is $\quad\quad \lambda = \dfrac{h}{nh/2\pi \, r_n}$

or $\quad\quad\quad\quad\quad\quad\quad\quad\quad\quad \lambda = \dfrac{2\pi \, r_n}{n}$ **[1]**

Note that this result says that $2\pi \, r_n = n\lambda$, or the allowed orbits are those whose circumference is an integral multiple of the de Broglie wavelength of the electron.

In the Bohr model, the radii of the allowed orbits in hydrogen are

$$r_n = n^2 a_0 \quad\quad \text{where} \quad\quad a_0 = 0.052\,9 \text{ nm} = 5.29 \times 10^{-11} \text{ m}$$

is the Bohr radius (the radius of the innermost orbit). Hence, Equation [1] becomes

$$\lambda = \frac{2\pi \, n^2 a_0}{n} = n(2\pi \, a_0) = n(3.32 \times 10^{-10} \text{ m})$$

The de Broglie wavelength of the electron in the third excited state (for which $n = 4$) in the hydrogen atom is then:

$$\lambda = 4(3.32 \times 10^{-10} \text{ m}) = 1.33 \times 10^{-9} \text{ m} = 1.33 \text{ nm} \quad\quad\quad \Diamond$$

35.   The ρ-meson has a charge of $-e$, a spin quantum number of 1, and a mass 1507 times that of the electron. If the electrons in atoms were replaced by ρ-mesons, list the possible sets of quantum numbers for ρ-mesons in the $3d$ subshell.

Solution

The $3d$ subshell has $n = 3$ and $\ell = 2$. Also, for ρ-mesons, the spin quantum number is $s = 1$. The orbital magnetic quantum number, m_ℓ, may vary from $-\ell$ to $+\ell$ in integer steps (a total of $2\ell + 1 = 5$ possible values). The spin magnetic quantum number varies from $-s$ to $+s$ in integer steps (a total of $2s + 1 = 3$ possible values). Thus, there are 15 possible quantum states for ρ-mesons in the $3d$ subshell. These states are summarized in the table given below:

n	3	3	3	3	3	3	3	3	3	3	3	3	3	3	3
ℓ	2	2	2	2	2	2	2	2	2	2	2	2	2	2	2
m_ℓ	+2	+2	+2	+1	+1	+1	0	0	0	−1	−1	−1	−2	−2	−2
m_s	+1	0	−1	+1	0	−1	+1	0	−1	+1	0	−1	+1	0	−1

◊

37. Two electrons in the same atom have $n=3$ and $\ell=1$. (a) List the quantum numbers for the possible states of the atom. (b) How many states would be possible if the exclusion principle did not apply to the atom?

Solution

(a) The Pauli exclusion principle states that no two electrons in the same atom can have identical sets of quantum numbers $(n, \ell, m_\ell,$ and $m_s)$. For states having given values of n and ℓ, the orbital magnetic quantum number, m_ℓ, may vary from $-\ell$ to $+\ell$ in integer steps. For each of these $(2\ell+1)$ possible values of m_ℓ, the spin magnetic quantum number m_s may take on either of two values, $\pm 1/2$. When $n=3$ and $\ell=1$, there are 6 possible states the first electron could be in. For each of these states, there are 5 other states the second electron could be in without violating the exclusion principle, giving a total of $6\times 5 = 30$ possible combinations of quantum numbers for the two electrons. These combinations are listed below. ◊

For both electrons, $n=3$ and $\ell=1$. The other quantum numbers are:

	m_ℓ	m_s	m_ℓ	m_s	m_ℓ	m_s	m_ℓ	m_s	m_ℓ	m_s	m_ℓ	m_s
Electron #1	+1	$+\frac{1}{2}$	+1	$+\frac{1}{2}$	+1	$+\frac{1}{2}$	+1	$-\frac{1}{2}$	+1	$-\frac{1}{2}$	+1	$-\frac{1}{2}$
Electron #2	+1	$-\frac{1}{2}$	0	$\pm\frac{1}{2}$	-1	$\pm\frac{1}{2}$	+1	$+\frac{1}{2}$	0	$\pm\frac{1}{2}$	-1	$\pm\frac{1}{2}$

	m_ℓ	m_s	m_ℓ	m_s	m_ℓ	m_s	m_ℓ	m_s	m_ℓ	m_s	m_ℓ	m_s
Electron #1	0	$+\frac{1}{2}$	0	$+\frac{1}{2}$	0	$+\frac{1}{2}$	0	$-\frac{1}{2}$	0	$-\frac{1}{2}$	0	$-\frac{1}{2}$
Electron #2	+1	$\pm\frac{1}{2}$	0	$-\frac{1}{2}$	-1	$\pm\frac{1}{2}$	+1	$\pm\frac{1}{2}$	0	$+\frac{1}{2}$	-1	$\pm\frac{1}{2}$

	m_ℓ	m_s	m_ℓ	m_s	m_ℓ	m_s	m_ℓ	m_s	m_ℓ	m_s	m_ℓ	m_s
Electron #1	-1	$+\frac{1}{2}$	-1	$+\frac{1}{2}$	-1	$+\frac{1}{2}$	-1	$-\frac{1}{2}$	-1	$-\frac{1}{2}$	-1	$-\frac{1}{2}$
Electron #2	+1	$\pm\frac{1}{2}$	0	$\pm\frac{1}{2}$	-1	$-\frac{1}{2}$	+1	$\pm\frac{1}{2}$	0	$\pm\frac{1}{2}$	-1	$+\frac{1}{2}$

(b) If the electrons did not obey the exclusion principle, there would be 36 possible combinations of quantum numbers for these two electrons. For each of the 6 possible states described above for the first electron, the second electron would have the choice of any of the 6 states (since it would be allowed to duplicate the state of the first electron). This would give a total of $6\times 6 = 36$ possible combinations. ◊

43. The K series of the discrete spectrum of tungsten contains wavelengths of 0.018 5 nm, 0.020 9 nm, and 0.021 5 nm. The K-shell ionization energy is 69.5 keV. Determine the ionization energies of the L, M, and N shells. Sketch the transitions that produce the above wavelengths.

Solution

Characteristic x-rays of the K series are produced when an electron in an upper energy level of a many-electron atom drops down to fill a vacancy in the K ($n = 1$) shell. Assuming that the given wavelengths are the three longest wavelengths in the K series of tungsten, the transitions that produce them are from the N ($n = 4$) shell to the K shell, from the M ($n = 3$) shell to the K shell, and from the L ($n = 2$) shell to the K shell, respectively. These transitions are shown in the energy level diagram where $\lambda_1 = 0.018\ 5$ nm, $\lambda_2 = 0.020\ 9$ nm, and $\lambda_3 = 0.021\ 5$ nm. If the ionization energy of the K shell is 69.5 keV, then $E_K = -69.5$ keV.

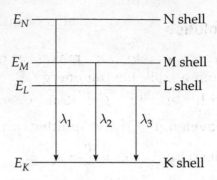

The photon produced in a transition has energy

$$E_\gamma = \frac{hc}{\lambda} = E_i - E_f$$

so the energy of the initial shell is

$$E_i = E_f + \frac{hc}{\lambda}$$

Thus, $E_N = E_K + \dfrac{hc}{\lambda_1} = -69.5\ \text{keV} + \dfrac{\left(6.63 \times 10^{-34}\ \text{J·s}\right)\left(3.00 \times 10^8\ \text{m/s}\right)}{0.018\ 5 \times 10^{-9}\ \text{m}} \left(\dfrac{1\ \text{keV}}{1.60 \times 10^{-16}\ \text{J}}\right)$

or $\quad E_N = -2.30$ keV

Similarly, $\quad E_M = E_K + \dfrac{hc}{\lambda_2} = -10.0\ \text{keV}$ and $E_L = E_K + \dfrac{hc}{\lambda_3} = -11.7$ keV

Thus, the ionization energies of the L, M, and N shells are:

11.7 keV, 10.0 keV, and 2.30 keV respectively ◊

47. The Lyman series for a (new?) one-electron atom is observed in a distant galaxy. The wavelengths of the first four lines and the short-wavelength limit of this Lyman series are given by the energy-level diagram in Figure P28.47. Based on this information, calculate (a) the energies of the ground state and first four excited states for this one-electron atom and (b) the longest-wavelength (alpha) line and the short-wavelength series limit in the Balmer series for this atom.

Solution

When an electron makes a transition from a state having energy E_i to a state with energy E_f, the energy and wavelength of the emitted photon are given by

$$E_\gamma = \frac{hc}{\lambda} = E_i - E_f$$

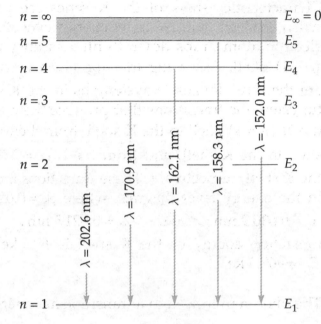

Figure P28.47

(a) The short-wavelength limit of the Lyman series involves the transition $(n = \infty \rightarrow n = 1)$, and from the figure, we see that when $E_i = E_\infty = 0$ and $E_f = E_1$, $\lambda = 152.0$ nm. Thus, the energy of the ground state in this atom is

$$E_1 = E_\infty - \frac{hc}{\lambda} = 0 - \frac{\left(6.63 \times 10^{-34}\ \text{J} \cdot \text{s}\right)\left(3.00 \times 10^{8}\ \text{m/s}\right)}{152.0 \times 10^{-9}\ \text{m}} \left(\frac{1\ \text{eV}}{1.60 \times 10^{-19}\ \text{J}}\right) = -8.18\ \text{eV} \qquad \Diamond$$

For the transition producing $\lambda = 202.6$ nm, $E_i = E_2$ and $E_f = E_1$. Thus, the energy of the first excited state is

$$E_2 = E_1 + \frac{hc}{\lambda} = -8.18\ \text{eV} + \frac{\left(6.63 \times 10^{-34}\ \text{J} \cdot \text{s}\right)\left(3.00 \times 10^{8}\ \text{m/s}\right)}{\left(202.6 \times 10^{-9}\ \text{m}\right)\left(1.60 \times 10^{-19}\ \text{J/eV}\right)} = -2.04\ \text{eV} \qquad \Diamond$$

Similarly, we find the energies of the next three excited states as

$$E_3 = E_1 + \frac{hc}{170.9\ \text{nm}} = -0.904\ \text{eV} \qquad\qquad E_4 = E_1 + \frac{hc}{162.1\ \text{nm}} = -0.510\ \text{eV} \qquad \Diamond$$

and $\quad E_5 = E_1 + \dfrac{hc}{158.3\ \text{nm}} = -0.325\ \text{eV} \qquad\qquad\qquad\qquad\qquad\qquad\qquad\qquad \Diamond$

(b)   The Balmer series involve transitions that terminate on the $n = 2$ level. For the longest wavelength (lowest energy photon) in this series, $n_i = 3$. Thus,

$$\lambda_\alpha = \frac{hc}{E_\gamma} = \frac{hc}{E_3 - E_2} = \frac{\left(6.63 \times 10^{-34} \text{ J} \cdot \text{s}\right)\left(3.00 \times 10^8 \text{ m/s}\right)}{\left[-0.904 \text{ eV} - \left(-2.04 \text{ eV}\right)\right]\left(1.60 \times 10^{-19} \text{ J/eV}\right)}$$

which gives the longest wavelength in this Balmer series as

$$\lambda_\alpha = 1.09 \times 10^{-6} \text{ m}\left(\frac{1 \text{ nm}}{10^{-9} \text{ m}}\right) = 1.09 \times 10^3 \text{ nm} \qquad \lozenge$$

For the short-wavelength limit (highest energy photon) in this series, $n_i = \infty$ and $n_f = 2$. This gives

$$\lambda_{min} = \frac{hc}{E_\infty - E_2} = \frac{\left(6.63 \times 10^{-34} \text{ J} \cdot \text{s}\right)\left(3.00 \times 10^8 \text{ m/s}\right)}{\left[0 - \left(-2.04 \text{ eV}\right)\right]\left(1.60 \times 10^{-19} \text{ J/eV}\right)}\left(\frac{1 \text{ nm}}{10^{-9} \text{ m}}\right) = 609 \text{ nm} \qquad \lozenge$$

51. A laser used in eye surgery emits a 3.00-mJ pulse in 1.00 ns, focused to a spot 30.0 μm in diameter on the retina. (a) Find (in SI units) the power per unit area at the retina. (This quantity is called the *irradiance*.) (b) What energy is delivered per pulse to an area of molecular size—say, a circular area 0.600 nm in diameter?

Solution

(a) The average power delivered to the retina during the laser pulse is

$$\mathcal{P}_{av} = \frac{\Delta E}{\Delta t} = \frac{3.00 \times 10^{-3} \text{ J}}{1.00 \times 10^{-9} \text{ s}} = 3.00 \times 10^{6} \text{ J/s} = 3.00 \times 10^{6} \text{ W}$$

This power is focused onto an area of

$$A = \frac{\pi d^2}{4} = \frac{\pi \left(30.0 \times 10^{-6} \text{ m}\right)^2}{4} = 2.25 \pi \times 10^{-10} \text{ m}^2$$

so the irradiance is

$$I = \frac{\mathcal{P}_{av}}{A} = \frac{3.00 \times 10^{6} \text{ W}}{2.25 \pi \times 10^{-10} \text{ m}^2} = 4.24 \times 10^{15} \text{ W/m}^2 \qquad \Diamond$$

(b) The energy delivered to a circular area 0.600 nm in diameter within the illuminated portion of the retina is

$$E = (IA)\Delta t = \left[I \left(\frac{\pi d^2}{4} \right) \right] \Delta t$$

or $E = \left[\left(4.24 \times 10^{15} \text{ } \frac{\text{W}}{\text{m}^2} \right) \frac{\pi \left(0.600 \times 10^{-9} \text{ m} \right)^2}{4} \right] \left(1.00 \times 10^{-9} \text{ s} \right) = 1.20 \times 10^{-12} \text{ J} \qquad \Diamond$

57.   In this problem you will estimate the classical lifetime of the hydrogen atom. An accelerating charge loses electromagnetic energy at a rate given by $\mathcal{P} = -2k_e q^2 a^2 / (3c^3)$, where k_e is the Coulomb constant, q is the charge of the particle, a is its acceleration, and c is the speed of light in a vacuum. Assume that the electron is one Bohr radius (0.052 9 nm) from the center of the hydrogen atom. (a) Determine its acceleration. (b) Show that $\mathcal{P}$ has units of energy per unit time and determine the rate of energy loss. (c) Calculate the kinetic energy of the electron and determine how long it will take for all of this energy to be converted into electromagnetic waves, assuming that the rate calculated in part (b) remains constant throughout the electron's motion.

Solution

(a)   The force accelerating the electron is the electrical attraction between it and the proton. Thus, from Newton's second law, the acceleration is

$$a = \frac{F}{m_e} = \frac{k_e e^2 / r^2}{m_e} = \frac{(8.99 \times 10^9 \ \text{N} \cdot \text{m}^2 / \text{C}^2)(1.60 \times 10^{-19} \ \text{C})^2}{(9.11 \times 10^{-31} \ \text{kg})(0.052\,9 \times 10^{-9} \ \text{m})^2} = 9.03 \times 10^{22} \ \text{m/s}^2 \qquad \Diamond$$

(b) $$[\mathcal{P}] = \frac{[k_e][q^2][a^2]}{[c^3]} = \frac{(\text{N} \cdot \text{m}^2 / \text{C}^2)(\text{C}^2)(\text{m}^2 / \text{s}^4)}{(\text{m}^3 / \text{s}^3)} = \frac{\text{N} \cdot \text{m}}{\text{s}} = \frac{\text{J}}{\text{s}} = \frac{[\text{energy}]}{[\text{time}]} \qquad \Diamond$$

The rate of energy loss (that is, the radiated power) is

$$|\mathcal{P}| = \frac{2k_e q^2 a^2}{3c^3} = \frac{2(8.99 \times 10^9 \ \text{N} \cdot \text{m}^2 / \text{C}^2)(1.60 \times 10^{-19} \ \text{C})^2 (9.03 \times 10^{22} \ \text{m/s}^2)^2}{3(3.00 \times 10^8 \ \text{m/s})^3}$$

which yields $|\mathcal{P}| = 4.63 \times 10^{-8} \ \text{J/s} = 4.63 \times 10^{-8} \ \text{W}$ \qquad $\Diamond$

(c)   The centripetal acceleration of the electron is $v^2 / r = a = 9.03 \times 10^{22} \ \text{m/s}^2$

giving, $v^2 = ra = (0.052\,9 \times 10^{-9} \ \text{m})(9.03 \times 10^{22} \ \text{m/s}^2) = 4.78 \times 10^{12} \ \text{m}^2 / \text{s}^2$

The kinetic energy of the orbiting electron is then

$$KE = \frac{1}{2} m_e v^2 = \frac{1}{2}(9.11 \times 10^{-31} \ \text{kg})(4.78 \times 10^{12} \ \text{m}^2 / \text{s}^2) = 2.18 \times 10^{-18} \ \text{J}$$

At the rate of energy loss found in Part (b), the time required to radiate all of this energy (and hence an estimate of the lifetime of the atom according to classical theory) is

$$\Delta t = \frac{KE}{|\mathcal{P}|} = \frac{2.18 \times 10^{-18} \ \text{J}}{4.63 \times 10^{-8} \ \text{J/s}} = 4.70 \times 10^{-11} \ \text{s} \qquad \text{or} \qquad \Delta t \sim 10^{-11} \ \text{s} \qquad \Diamond$$

Chapter 29

Nuclear Physics

NOTES FROM SELECTED CHAPTER SECTIONS

29.1   Some Properties of Nuclei
Important terms in the description of nuclear properties include:

- The **atomic number, Z,** — equals the number of protons in the nucleus.

- The **neutron number, N,** — equals the number of neutrons in the nucleus.

- The **mass number, A,** — equals the number of nucleons (neutrons plus protons) in the nucleus.

Nuclei can be represented symbolically as $^A_Z X$, where X designates the chemical symbol for a specific chemical element.

The nuclei of all atoms of a particular element contain the same number of protons but may contain different numbers of neutrons. Nuclei that are related in this way are called **isotopes. The isotopes of an element have the same Z value but different N and A values.**

The **unified mass unit, u, is defined such that the mass of the isotope ^{12}C is exactly 12 u** (1 u = $1.660\,559 \times 10^{-27}$ kg).

Based on nuclear scattering and other experiments, it is possible to conclude that:

- Most nuclei are approximately spherical, and all have nearly the same density.

- Nucleons combine to form a nucleus as though they were tightly packed spheres.

- The stability of nuclei is due to a short-range attractive nuclear force.

- The nuclear force is approximately the same for an interaction between any pair of nucleons (p-p, n-n, or p-n).

29.2 Binding Energy

The total mass of a nucleus is always less than the sum of the masses of its individual nucleons. This difference in mass is the origin of the nuclear **binding energy** and represents the energy that would be required to separate the nucleus into neutrons and protons.

29.3 Radioactivity

The decay of radioactive substances can be accompanied by the emission of three forms of radiation that vary in ability to penetrate shielding materials:

- **alpha** (α) **particles** — Helium nuclei (^4_2He) which can barely penetrate a sheet of paper.

- **beta** (β) **particles** — Either electrons (e^-) or positrons (e^+) able to penetrate a few millimeters of aluminum.

- **gamma** (γ) **rays** — High-energy photons (γ) capable of penetrating several centimeters of lead.

A positron is the antiparticle of the electron; it has a mass equal to m and a charge of $+q_e$.

The decay rate, or activity, R, of a sample of a radioactive substance is defined as the number of decays occurring per second. The half life ($T_{1/2}$) of a radioactive substance is the time during which half of an initial number of radioactive nuclei in a sample will decay. The customary unit of radioactivity is the Curie (Ci); the SI unit of activity is the **becquerel** (Bq), where 1 Bq = 1 decay/s and $1\,\text{Ci} = 3.7 \times 10^{10}\,\text{Bq}$.

29.4 The Decay Processes

The overall decay process can be represented in equation form as:

$$\underset{\substack{\text{Parent}\\\text{nucleus}}}{X} \rightarrow \underset{\substack{\text{Daughter}\\\text{nucleus}}}{Y} + \overset{\text{emitted}}{\text{radiation}}$$

In an equation representing a specific radioactive decay process, the sum of the mass numbers on each side of the equation must be equal, and the sum of the atomic numbers on each side of the equation must be equal.

The neutrino and antineutrino shown in the beta decay processes in the table below are required for conservation of energy and momentum.

The neutrino has the following properties:

- zero electric charge

- rest mass much smaller than that of the electron (recent experiments suggest that the neutrino mass may not be zero)

- spin of $\frac{1}{2}$ (satisfying the law of conservation of angular momentum)

- very weak interaction with matter (therefore very difficult to detect)

General characteristics of alpha, beta, and gamma decay are summarized below.

Decay	can be written as	Process
Alpha decay	$^{A}_{Z}X \rightarrow \, ^{A-4}_{Z-2}Y + \, ^{4}_{2}He$	The parent nucleus emits an alpha particle and loses two protons and two neutrons.
Beta decay	$^{A}_{Z}X \rightarrow \, ^{A}_{Z+1}Y + e^{-} + \bar{\nu}$ (electron emission) $^{A}_{Z}X \rightarrow \, ^{A}_{Z-1}Y + e^{+} + \nu$ (positron emission)	A nucleus can undergo beta decay in two ways. The parent nucleus can emit an electron (e^{-}) and an antineutrino ($\bar{\nu}$) or emit a positron (e^{+}) and a neutrino (ν).
Gamma decay	$^{A}_{Z}X^{*} \rightarrow \, ^{A}_{Z}X + \gamma$ * parent nucleus in an excited state	A nucleus in an excited state decays to a lower energy state (often the ground state) and emits a gamma ray.

29.5   Natural Radioactivity

Natural radioactivity is the decay of radioactive nuclei found in nature.

Artificial radioactivity is the decay of nuclei produced in the laboratory through nuclear reactions.

Most naturally decaying radionuclides are members of a **decay series**.

The Uranium Series, Actinium Series, and Thorium Series each begin with a long-half life naturally occurring radioisotope and, following a sequence of decays, ends in a stable isotope. A fourth series, the Neptunium Series, begins with the decay of an artificially produced radionuclide.

29.6 Nuclear Reactions

Nuclear reactions are events in which collisions between nuclei change the structure or properties of nuclei. The *Q-value* is the energy required to balance the overall reaction equation.

Exothermic reactions release energy and have positive *Q*-values. The *Q*-values of endothermic reactions are negative, and the minimum energy of the incoming particle for which an endothermic reaction will occur is called the **threshold energy.**

29.7 Medical Applications of Radiation

Radiation damage to cells in biological organisms is primarily due to ionizing events and can be separated into two categories:

- **Somatic damage** is radiation damage to cells other than reproductive cells.

- **Genetic damage** affects only reproductive cells.

Several units are used to quantify radiation exposure and dose:

- **Roengten (R)**; the amount of ionizing radiation that will produce 2.08×10^9 ion pairs in 1 cm^3 of air under standard conditions.

- **rad**; the amount of radiation that deposits 10^{-2} J of energy into 1 kg of absorbing material.

- **RBE** (relative biological effectiveness) factor; the number of rad of either x-ray or gamma radiation that will produce the same degree of biological damage as the actual radiation being used (e.g. electrons, protons, neutrons, heavy ions, etc.).

- **rem** (roentgen equivalent in man); the product of the dose measured in rad and the RBE factor. One rem of any two types of radiation will produce the same degree of biological damage.
  Dose in rem = dose in rad × RBE.

EQUATIONS AND CONCEPTS

The **average nuclear radius** is proportional to the cube root of the mass number (or total number of nucleons). This means that the volume is proportional to A, nuclei are approximately spherical in shape, and all nuclei have nearly the same density.

$$r = r_0 A^{1/3}$$ (29.1)
$$r_0 = 1.2 \text{ fm}$$

The **fermi (fm)** is a convenient unit of length to use in stating nuclear dimensions.

$$1 \text{ fm} = 10^{-15} \text{ m}$$

The **decay constant**, λ, is characteristic of a particular isotope. The number of radioactive nuclei in a given sample which undergo decay during a time interval Δt is proportional to the number of nuclei present.

$$\Delta N = -\lambda N \Delta t$$ (29.2)

The decay rate or activity, R, of a sample of radioactive nuclei is defined as the number of decays per second.

$$R = \left| \frac{\Delta N}{\Delta t} \right| = \lambda N$$ (29.3)

The **decay curve** is a plot of, N, the number of nuclei remaining in a sample versus the time of decay. The number of nuclei remaining in a radioactive substance, decreases exponentially with time. N_0 is the number of nuclei in the sample at $t = 0$. In Equation (29.4b), n is the number of half-lives that have elapsed.

$$N = N_0 e^{-\lambda t}$$ (29.4a)
$$N = N_0 \left(\frac{1}{2} \right)^n$$ (29.4b)

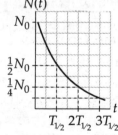

The **half-life of a radioactive substance** is the time required for half of the radioactive nuclei remaining in a sample of the substance to decay.

$$T_{1/2} = \frac{\ln 2}{\lambda} = \frac{0.693}{\lambda}$$ (29.5)

Activity, R, can be expressed in units of **becquerels or curies**.

$$1 \text{ Ci} \equiv 3.7 \times 10^{10} \text{ decays/s} \qquad (29.6)$$

$$1 \text{ Bq} = 1 \text{ decay/s} \qquad (29.7)$$

In an **alpha decay process**, the parent nucleus loses two neutrons and two protons. Alpha decay can be represented symbolically in terms of the parent nucleus, $_{Z}^{A}X$, and the daughter nucleus, $_{Z-2}^{A-4}Y$. An example is the decay of $_{92}^{238}U$ as shown in Equation 29.9.

$$_{Z}^{A}X \rightarrow {}_{Z-2}^{A-4}Y + {}_{2}^{4}He \qquad (29.8)$$

$$_{92}^{238}U \rightarrow {}_{90}^{234}Th + {}_{2}^{4}He \qquad (29.9)$$

For $_{92}^{238}U$, $T_{1/2} = 4.47 \times 10^{9}$ yr

In order for alpha emission to occur, the mass of the parent nucleus must be greater than the combined mass of the daughter nucleus and the emitted alpha particle. The mass difference is converted into energy and appears as kinetic energy shared (unequally) by the alpha particle and the daughter nucleus.

Mass (energy) requirement for alpha decay

The **beta decay process** is accompanied by emission of a third particle that is required for conservation of energy and momentum: antineutrino ($\bar{\nu}$) with e^{-} and neutrino (ν) with e^{+}.

$$_{Z}^{A}X \rightarrow {}_{Z+1}^{A}Y + e^{-} + \bar{\nu}$$

$$_{Z}^{A}X \rightarrow {}_{Z-1}^{A}Y + e^{+} + \nu$$

Examples of electron and positron emission are shown in Equations 29.15 and 29.16.

$$_{6}^{14}C \rightarrow {}_{7}^{14}N + e^{-} + \bar{\nu} \qquad (29.15)$$

$$_{7}^{12}N \rightarrow {}_{6}^{12}C + e^{+} + \nu \qquad (29.16)$$

The electron that is emitted in beta decay is created within the parent nucleus by a process which can be represented as a neutron being transformed into a proton and an electron.

$$_{0}^{1}n \rightarrow {}_{1}^{1}p + e^{-} \qquad (29.14)$$

Gamma ray emission often follows alpha or beta decay when the daughter nucleus is left in an **excited energy state**. The nucleus returns to the ground state by emission of one or more photons. The mass number and the atomic number do not change as a result of gamma ray emission. Equations 29.17 and 29.18 show a typical sequence of of events in which the * symbol is used to identify a nucleus in an excited state.

$$^{12}_{5}B \rightarrow ^{12}_{6}C* + e^- + \bar{\nu} \tag{29.17}$$

$$^{12}_{6}C* \rightarrow ^{12}_{6}C + \gamma \tag{29.18}$$

Artificial transmutations (nuclear reactions) can be induced by bombarding stable target nuclei with energetic particles. Equation 29.20 shows the equation for the first nuclear reaction produced in the laboratory. **Note that both reactants and both products in the reaction are stable isotopes.**

$$^{4}_{2}He + ^{14}_{7}N \rightarrow ^{17}_{8}O + ^{1}_{1}H \tag{29.21}$$

The Q value of a nuclear reaction is the quantity of energy required to balance the equation representing the reaction. **Q is positive for exothermic reactions and negative for endothermic reactions.**

Energy conservation in nuclear reactions: Q values

$$Q = KE_{products} - KE_{reactants}$$

$$Q = (m_{reactants} - m_{products})c^2$$

A **threshold energy** (minimum kinetic energy of the incident particle) is characteristic of endothermic reactions. This minimun value of kinetic energy of the incident particle is required in order for energy and momentum to be conserved.

$$KE_{min} = \left(1 + \frac{m}{M}\right)|Q| \tag{29.24}$$

m = mass of incident particle

M = mass of target particle

SUGGESTIONS, SKILLS, AND STRATEGIES

The rest energy of a particle is given by $E = mc^2$. It is therefore often convenient to express the unified mass unit in terms of its energy equivalent, $1\,u = 1.660\,559 \times 10^{-27}$ kg or $1\,u = 931.50\,\text{MeV}/c^2$. When masses are expressed in units of u, energy values are then $E = m(931.50\,\text{MeV}/u)$.

Equation 29.4a can be solved for the particular time t after which the number of remaining nuclei will be some specified fraction (N/N_0) of the original number N_0. This can be done by taking the natural log of each side of Equation 29.4a to find

$$t = \left(\frac{1}{\lambda}\right)\ln\left(\frac{N_0}{N}\right)$$

Then substitute for N the specified fraction of N_0.

REVIEW CHECKLIST

▷ Identify each of the components of radiation that are emitted by nuclei through natural radioactive decay and describe the basic properties of each. Write out typical equations to illustrate the processes of transmutation by alpha and beta decay.

▷ State and apply to the solution of related problems, the formula which expresses decay rate as a function of decay constant and number of radioactive nuclei. Also apply the exponential formula which expresses the number of radioactive nuclei remaining as a function of elapsed time, decay constant, or half-life, and the initial number of nuclei.

▷ Calculate the Q value of a given nuclear reaction and determine the threshold energy of an endothermic reaction.

▷ Be familiar with the various units used to express quantities of radiation dose.

▷ Given the reactants and one of the products of a nuclear reaction, be able to balance the reaction equation and identify the remaining product nuclide.

SOLUTIONS TO SELECTED END-OF-CHAPTER PROBLEMS

5. An alpha particle $(Z = 2,\ \text{mass } 6.64 \times 10^{-27}\ \text{kg})$ approaches to within 1.00×10^{-14} m of a carbon nucleus $(Z = 6)$. What are (a) the maximum Coulomb force on the alpha particle, (b) the acceleration of the alpha particle at this time, and (c) the potential energy of the alpha particle at the same time?

Solution

(a) The repulsive electrical force exerted on the alpha particle by the carbon nucleus has a magnitude of

$$F = \frac{k_e q_1 q_2}{r^2} = \frac{k_e (2e)(6e)}{r^2} = \frac{12 k_e e^2}{r^2}$$

The maximum of this force occurs at the distance of closest approach, so

$$F_{max} = \frac{12 k_e e^2}{r_{min}^2} = \frac{12 \left(8.99 \times 10^9\ \text{N} \cdot \text{m}^2/\text{C}^2\right)\left(1.60 \times 10^{-19}\ \text{C}\right)^2}{\left(1.00 \times 10^{-14}\ \text{m}\right)^2} = 27.6\ \text{N} \qquad \Diamond$$

(b) At the point of closest approach, the acceleration of the alpha particle is

$$a_{max} = \frac{F_{max}}{m_\alpha} = \frac{27.6\ \text{N}}{6.64 \times 10^{-27}\ \text{kg}} = 4.16 \times 10^{27}\ \text{m/s}^2 \qquad \Diamond$$

(c) The electrical potential energy of a point charge q_1 at distance r from a second point charge q_2 is

$$PE = q_1 V = q_1 \left(\frac{k_e q_2}{r} \right) = \frac{k_e q_1 q_2}{r}$$

Thus, the potential energy of the alpha particle at the point of its closest approach to the carbon nucleus is

$$PE_{max} = \frac{k_e (2e)(6e)}{r_{min}} = \frac{12 \left(8.99 \times 10^9\ \text{N} \cdot \text{m}^2/\text{C}^2\right)\left(1.60 \times 10^{-19}\ \text{C}\right)^2}{1.00 \times 10^{-14}\ \text{m}}$$

or $\qquad PE_{max} = 2.76 \times 10^{-13}\ \text{J} \left(\frac{1\ \text{MeV}}{1.60 \times 10^{-13}\ \text{J}} \right) = 1.73\ \text{MeV} \qquad \Diamond$

11. A pair of nuclei for which $Z_1 = N_2$ and $Z_2 = N_1$ are called *mirror isobars*. (The atomic and neutron numbers are interchangeable.) Binding-energy measurements on such pairs can be used to obtain evidence of the charge independence of nuclear forces. Charge independence means that the proton-proton, proton-neutron, and neutron-neutron forces are approximately equal. Calculate the difference in binding energy for the two mirror nuclei $^{15}_{8}\text{O}$ and $^{15}_{7}\text{N}$.

Solution

The binding energy of a nucleus $^A_Z X$ is given by $E_b = (\Delta m)c^2$, where the mass deficit is

$$\Delta m = \left[Zm_{^1_1\text{H}} + (A - Z)m_n \right] - m_{^A_Z X}$$

Here, Z is the atomic number (number of protons in the nucleus), $(A - Z)$ is the number of neutrons, m_n is the mass of a neutron, $m_{^1_1\text{H}}$ is the mass of a neutral hydrogen atom, and $m_{^A_Z X}$ is the mass of a neutral atom containing the nucleus of interest. We may use atomic masses from Appendix B of the textbook in this calculation since the included masses of Z electrons will cancel in the subtraction process.

For $^{15}_{8}\text{O}$, $\Delta m = 8(1.007\,825\ \text{u}) + 7(1.008\,665\ \text{u}) - 15.003\,065\ \text{u} = 0.120\,190\ \text{u}$

and the binding energy is $E_b = 0.120\,190\ \text{u} \cdot c^2$. The energy equivalent of the atomic mass unit is $1\ \text{u} \cdot c^2 = 931.5\ \text{MeV}$, so the binding energy of $^{15}_{8}\text{O}$ is

$$E_b = (0.120\,190\ \text{u})(931.5\ \text{MeV/u}) = 112.0\ \text{MeV}$$

For $^{15}_{7}\text{N}$, $\Delta m = 7(1.007\,825\ \text{u}) + 8(1.008\,665\ \text{u}) - 15.000\,108\ \text{u} = 0.123\,987\ \text{u}$

and $E_b = (0.123\,987\ \text{u})(931.5\ \text{MeV/u}) = 115.5\ \text{MeV}$

Thus, the difference in the binding energies of these mirror isobars is

$$\Delta E_b = E_b \big|_{^{15}\text{N}} - E_b \big|_{^{15}\text{O}} = 115.5\ \text{MeV} - 112.0\ \text{MeV} = 3.5\ \text{MeV} \qquad \Diamond$$

15. The half-life of an isotope of phosphorus is 14 days. If a sample contains 3.0×10^{16} such nuclei, determine its activity. Express your answer in curies.

Solution

The activity, R, of a radioactive sample is directly proportional to the number of radioactive nuclei in that sample. That is,

$$R = \lambda N$$

where N is the number of nuclei in the sample, and λ is the decay constant of that species of unstable nuclei.

If R_0 is the activity of the sample at time $t = 0$, the activity at any later time is

$$R = R_0 e^{-\lambda t}$$

When $t = T_{1/2}$ (the half-life for this type nucleus), the activity is $R = R_0/2$, so

$$\frac{1}{2} R_0 = R_0 e^{-\lambda T_{1/2}} \qquad \text{or} \qquad e^{\lambda T_{1/2}} = 2$$

This yields $\quad \lambda T_{1/2} = \ln 2 \qquad$ or $\qquad \lambda = \dfrac{\ln 2}{T_{1/2}}$

Thus, if $T_{1/2} = 14$ d, the decay constant is $\quad \lambda = \dfrac{\ln 2}{14 \text{ d}}$

and the activity in curies $\left(1 \text{ Ci} = 3.7 \times 10^{10} \text{ decays per second}\right)$ of a sample containing $N = 3.0 \times 10^{16}$ radioactive nuclei of this type is

$$R = \lambda N = \left(\frac{\ln 2}{14 \text{ d}}\right)\left(3.0 \times 10^{16}\right)\left(\frac{1 \text{ d}}{86\ 400 \text{ s}}\right)\left(\frac{1 \text{ Ci}}{3.7 \times 10^{10} \text{ s}^{-1}}\right) = 0.46 \text{ Ci} \qquad \Diamond$$

21. Many smoke detectors use small quantities of the isotope ^{241}Am in their operation. The half-life of ^{241}Am is 432 yr. How long will it take for the activity of this material to decrease to 1.00×10^{-3} of the original activity?

Solution

The activity of a radioactive sample at time t is given by

$$R = R_0 e^{-\lambda t}$$

where R_0 is the original activity (at time $t = 0$) and λ is the decay constant for this type nuclei. The decay constant is related to the half-life by the expression

$$\lambda = \frac{\ln 2}{T_{1/2}}$$

Thus, the decay constant for ^{241}Am is $\qquad \lambda = \dfrac{\ln 2}{432 \text{ yr}}$

and when the activity has decreased to 1.00×10^{-3} of the original value,

$$\frac{R_0}{1\,000} = R_0 e^{-\lambda t} \qquad \text{or} \qquad e^{\lambda t} = 1\,000$$

Taking the natural logarithm of both sides of the last result yields

$$\lambda t = \ln(1\,000)$$

or $\qquad t = \dfrac{\ln(1\,000)}{\lambda} = \dfrac{\ln(1\,000)}{\ln 2 / 432 \text{ yr}} = (432 \text{ yr}) \dfrac{\ln(1\,000)}{\ln 2} = 4.31 \times 10^3 \text{ yr}$ $\qquad \Diamond$

29. The mass of ^{56}Fe is 55.9349 u, and the mass of ^{56}Co is 55.9399 u. Which isotope decays into the other and by what process?

Solution

Since the parent nucleus is giving up energy, and energy is equivalent to mass, the daughter nucleus must be less massive than the parent. Thus, the cobalt must decay into iron. ◊

To determine the decay process involved, realize that both charge and the total mass number must be conserved in the decay. Since both the parent and daughter nuclei in this decay have same mass number (56), the emitted particle must have a mass number of zero. The parent nucleus, $^{56}_{27}$Co, has 27 protons while the daughter, $^{56}_{26}$Fe, has 26 protons. Hence, the emitted particle must have a charge of $+e$. The particle that fits both of these criteria is the positron e^+. This must be positron emission (also called β^+ or e^+ decay). ◊

The decay equation is $^{56}_{27}$Co $\rightarrow$ $^{56}_{26}$Fe $+ e^+ + \nu$. Note that a neutral particle with zero mass (that is, the neutrino, ν) must be produced in this decay to conserve electron-lepton number which is discussed in the next chapter.

35. A wooden artifact is found in an ancient tomb. Its carbon-14 $\left(^{14}_{6}C\right)$ activity is measured to be 60.0% of that in a fresh sample of wood from the same region. Assuming the same amount of ^{14}C was initially present in the wood from which the artifact was made, determine the age of the artifact.

Solution

The activity of a radioactive sample is directly proportional to the number of radioactive nuclei contained in that sample. Thus, if the carbon-14 activity of the artifact is 60.0% that of an equal amount of fresh wood, the artifact must contain 60.0% of the number of carbon-14 nuclei found in the fresh sample. If it is assumed that the artifact originally contained the same number of ^{14}C nuclei as is found in a equal quantity of fresh wood, then we have

$$N = 0.600N_0$$

or, 60.0% of the ^{14}C nuclei originally in the wood of the artifact are still present in that sample.

Since $R = \lambda N$, we then have
$$R = \lambda(0.600N_0) = 0.600(\lambda N_0) = 0.600R_0$$

where R_0 is the activity the artifact had when its wood was freshly cut.

But
$$R = R_0 e^{-\lambda t} = R_0 e^{-\left(\ln 2/T_{1/2}\right)t}$$

which gives $\quad 0.600R_0 = R_0 e^{-\left(\ln 2/T_{1/2}\right)t} \qquad$ or $\qquad \ln(0.600) = -\left(\dfrac{\ln 2}{T_{1/2}}\right)t$

The half-life of ^{14}C is $T_{1/2} = 5730 \text{ yr}$, so the time that has passed since the wood in the artifact was part of a living tree is

$$t = -\left(\frac{\ln 0.600}{\ln 2}\right)T_{1/2} = -\left(\frac{\ln 0.600}{\ln 2}\right)(5730 \text{ yr}) = 4.22 \times 10^3 \text{ yr} \qquad \Diamond$$

41. (a) Suppose $^{10}_{5}\text{B}$ is struck by an alpha particle, releasing a proton and a product nucleus in the reaction. What is the product nucleus? (b) An alpha particle and a product nucleus are produced when $^{13}_{6}\text{C}$ is struck by a proton. What is the product nucleus?

Solution

In any nuclear reaction, the total number of nucleons (that is, the sum of the mass numbers of the particles involved) and the total charge (the sum of the atomic numbers of the involved particles) must have the same values after the reaction as before the reaction.

(a)   When $^{10}_{5}\text{B}$ is struck by an alpha particle $\left(^{4}_{2}\text{He}\right)$, releasing a proton $\left(^{1}_{1}\text{H}\right)$ and a product nucleus $\left(^{A}_{Z}X\right)$, the reaction equation is

$$^{10}_{5}\text{B} + {}^{4}_{2}\text{He} \rightarrow {}^{A}_{Z}X + {}^{1}_{1}\text{H}$$

Requiring that charge be conserved gives $5 + 2 = Z + 1$ or $Z = 6$

Thus, the product nucleus must be an isotope of carbon.

Requiring that number of nucleons be conserved gives

$$10 + 4 = A + 1 \qquad \text{or} \qquad A = 13$$

Therefore, the product nucleus is   carbon-13   or $^{13}_{6}\text{C}$ ◊

(b)   In the reaction $^{13}_{6}\text{C} + {}^{1}_{1}\text{H} \rightarrow {}^{A}_{Z}X + {}^{4}_{2}\text{He}$

Conservation of charge gives $6 + 1 = Z + 2$ or $Z = 5$

so the product nucleus is a Boron isotope.

Conserving number of nucleons gives $13 + 1 = A + 4$ or $A = 10$

Thus, if the reaction occurs, the product nucleus will be $^{10}_{5}\text{B}$ ◊

45. When ^{18}O is struck by a proton, ^{18}F and another particle are produced. (a) What is the other particle? (b) The reaction has a Q value of -2.453 MeV and the atomic mass of ^{18}O is $17.999\,160$ u. What is the atomic mass of ^{18}F?

Solution

(a) The reaction equation is

$$^{18}_{8}O + {}^{1}_{1}H \rightarrow {}^{18}_{9}F + {}^{A}_{Z}X$$

where $^{A}_{Z}X$ is the unknown particle

Requiring that charge be conserved gives $\qquad 8+1=9+Z$

or $\qquad Z=0$

Thus, the unknown must be neutral. Balancing the mass number on the left and right sides of the reaction equation gives

$$18+1=18+A$$

so $\qquad A=1$

Therefore, the unknown particle is neutral and has a mass number of 1.

This is a neutron $\qquad\qquad {}^{1}_{0}n$ ◊

(b) The Q value of a reaction is $\qquad Q=(\Delta m)c^2$

If $Q=-2.453$ MeV for this reaction,

the mass deficit $\qquad \Delta m = m_{^{18}_{8}O} + m_{^{1}_{1}H} - m_{^{18}_{9}F} - m_{^{1}_{0}n}$

must be $\qquad \Delta m = \dfrac{-2.453\text{ MeV}}{c^2}\left(\dfrac{1\text{ u}}{931.5\text{ MeV}/c^2}\right) = -0.002\,633$ u

Using atomic masses from Appendix B,

$$m_{^{18}_{9}F} = m_{^{18}_{8}O} + m_{^{1}_{1}H} - m_{^{1}_{0}n} - \Delta m$$

$$m_{^{18}_{9}F} = 17.999\,160\text{ u} + 1.007\,825\text{ u} - 1.008\,665\text{ u} + 0.002\,633\text{ u}$$

This gives the mass of ^{18}F, $\qquad m_{^{18}_{9}F} = 18.000\,953$ u ◊

51. A patient swallows a radiopharmaceutical tagged with phosphorus-32 $\left(^{32}_{15}P\right)$, a β^- emitter with a half-life of 14.3 days. The average kinetic energy of the emitted electrons is 700 keV. If the initial activity of the sample is 1.31 MBq, determine (a) the number of electrons emitted in a 10-day period, (b) the total energy deposited in the body during the 10 days, and (c) the absorbed dose if the electrons are completely absorbed in 100 g of tissue.

Solution

(a)   The decay constant for ^{32}P is:

$$\lambda = \frac{\ln 2}{T_{1/2}} = \frac{\ln 2}{14.3 \text{ days}} = 4.85 \times 10^{-2} \text{ days}^{-1}$$

Thus, the initial number of ^{32}P nuclei present is

$$N_0 = \frac{R_0}{\lambda} = \frac{1.31 \times 10^6 \text{ Bq}}{4.85 \times 10^{-2} \text{ days}^{-1}} \left(\frac{1 \text{ decay/s}}{1 \text{ Bq}}\right) \left(\frac{86\,400 \text{ s}}{1 \text{ day}}\right) = 2.335 \times 10^{12}$$

At time t, the number of ^{32}P nuclei remaining in the sample is $N = N_0 e^{-\lambda t}$. Hence, the number of decays that have occurred in the elapsed time is

$$\Delta N = N_0 - N = N_0 \left(1 - e^{-\lambda t}\right)$$

The number of decays occurring in this sample during the first 10.0 days is

$$\Delta N = N_0 \left[1 - e^{-\lambda(10.0 \text{ days})}\right] = 2.335 \times 10^{12} \left[1 - e^{-\left(4.85 \times 10^{-2} \text{ days}^{-1}\right)(10.0 \text{ days})}\right]$$

or $\Delta N = 8.97 \times 10^{11}$

Since one β^- particle (electron) is emitted per decay, the number of electrons emitted in 10.0 days is $\qquad 8.97 \times 10^{11}$ ◊

(b)   Each electron deposits its kinetic energy (an average of 700 keV per electron) in the body. The total energy deposited in the first 10.0 days is

$$E = \Delta N (700 \text{ keV}) = \left(8.97 \times 10^{11}\right)(700 \text{ keV}) \left(\frac{1.60 \times 10^{-16} \text{ J}}{1 \text{ keV}}\right) = 0.100 \text{ J} \qquad ◊$$

(c)   If the electrons are absorbed by 100 g of tissue, the energy deposited per unit mass is:

$$\text{dose} = \frac{E}{m} = \frac{1.00 \times 10^{-1} \text{ J}}{0.100 \text{ kg}} = 1.00 \text{ J/kg}$$

The rad is a dosage unit equal to 10^{-2} J/kg, so the absorbed dose in rad units is

$$\text{absorbed dose} = \left(1.00\ \frac{\text{J}}{\text{kg}}\right)\left(\frac{1\ \text{rad}}{10^{-2}\ \text{J/kg}}\right) = 100\ \text{rad} \qquad \Diamond$$

---

53.   A 200.0-mCi sample of a radioactive isotope is purchased by a medical supply house. If the sample has a half-life of 14.0 days, how long will it keep before its activity is reduced to 20.0 mCi?

Solution

If a radioactive sample has an activity R_0 at time $t = 0$, its activity at any time t later is given by

$$R = R_0 e^{-\lambda t}$$

where the decay constant λ is related to the half-life by the expression

$$\lambda = \frac{\ln 2}{T_{1/2}}$$

Thus, if the sample had an initial activity of $R_0 = 200.0$ mCi, we find the time when the activity will be $R = 20.0$ mCi $= 0.100\ R_0$ from

$$0.100 R_0 = R_0 e^{-\lambda t} \qquad \text{or} \qquad e^{\lambda t} = 10.0$$

Computing the natural logarithm of both sides of the last result gives

$$\lambda t = \ln(10.0) \qquad \text{or} \qquad t = \frac{\ln(10.0)}{\lambda} = \left[\frac{\ln(10.0)}{\ln 2}\right]T_{1/2}$$

Thus, if the sample has a half-life of $T_{1/2} = 14.0$ d,

$$t = \left[\frac{\ln(10.0)}{\ln 2}\right](14.0\ \text{d}) = 46.5\ \text{d} \qquad \Diamond$$

---

59. In a piece of rock from the Moon, the ^{87}Rb content is assayed to be 1.82×10^{10} atoms per gram of material, and the ^{87}Sr content is found to be 1.07×10^{9} atoms per gram. (The relevant decay is ^{87}Rb $\to$ ^{87}Sr $+ e^{-}$. The half-life of the decay is 4.8×10^{10} years.) (a) Determine the age of the rock. (b) Could the material in the rock actually be much older? What assumption is implicit in using the radioactive-dating method?

Solution

(a) If we assume all the ^{87}Sr nuclei present came from the decay of ^{87}Rb nuclei, the original number of ^{87}Rb nuclei present in a one gram sample of the rock must have been

$$N_0 = N_{^{87}\text{Rb}} + N_{^{87}\text{Sr}} = 1.82 \times 10^{10} + 1.07 \times 10^{9}$$

The current number of ^{87}Rb nuclei in this sample is

$$N = N_{^{87}\text{Rb}} = 1.82 \times 10^{10}$$

Then, from $N = N_0 e^{-\lambda t}$, the elapsed time (and hence, the age of the rock) is

$$t = -\frac{\ln\left(N/N_0\right)}{\lambda} = -\frac{T_{1/2}\ln\left(N/N_0\right)}{\ln 2}$$

$$t = -\frac{\left(4.8 \times 10^{10} \text{ yr}\right)\ln\left(\dfrac{1.82 \times 10^{10}}{1.82 \times 10^{10} + 1.07 \times 10^{9}}\right)}{\ln 2} = 4.0 \times 10^{9} \text{ yr} \qquad \Diamond$$

(b) The above calculation assumes that the rock originally contained no ^{87}Sr nuclei, and all the currently present ^{87}Sr nuclei came from the decay of ^{87}Rb nuclei. Thus, the result of this calculation represents **an upper limit** on the age of the rock. If the rock originally contained some ^{87}Sr, the number of ^{87}Rb decays (and hence, the time elapsed since the origin of the rock) is lower than that computed in part (a). $\qquad \Diamond$

65.   During the manufacture of a steel engine component, radioactive iron $\left(^{59}\text{Fe}\right)$ is included in the total mass of 0.20 kg. The component is placed in a test engine when the activity due to the isotope is 20.0 μCi. After a 1 000-h test period, oil is removed from the engine and is found to contain enough ^{59}Fe to produce 800 disintegrations/min per liter of oil. The total volume of oil in the engine is 6.5 L. Calculate the total mass worn from the engine component per hour of operation. (The half-life of ^{59}Fe is 45.1 days.)

Solution

The total activity of the iron at the end of the 1 000-h test will be

$$R = R_0 e^{-\lambda t} = R_0 e^{-t \cdot \ln 2 / T_{1/2}}$$

$$= \left(20.0 \ \mu\text{Ci}\right) e^{-\left[\left(10^3 \ \text{h}\right) \cdot \ln 2 / (45.1 \ \text{d})(24 \ \text{h/d})\right]} = 10.5 \ \mu\text{Ci}$$

The activity due to iron worn from the part and now present in the oil is

$$R_{\text{oil}} = \left(\frac{800 \ \text{min}^{-1}}{\text{L}}\right)\left(6.5 \ \text{L}\right) = 5.2 \times 10^3 \ \text{min}^{-1}\left(\frac{1 \ \text{min}}{60 \ \text{s}}\right)\left(\frac{1 \ \mu\text{Ci}}{3.7 \times 10^4 \ \text{s}^{-1}}\right) = 2.3 \times 10^{-3} \ \mu\text{Ci}$$

Therefore, the fraction of the total mass of iron worn away during the test was

$$\text{fraction} = \frac{R_{\text{oil}}}{R} = \frac{2.3 \times 10^{-3} \ \mu\text{Ci}}{10.5 \ \mu\text{Ci}} = 2.2 \times 10^{-4}$$

This represents a mass of

$$m_{\text{worn away}} = \left(\text{fraction}\right) \cdot \left(\text{total mass of iron}\right)$$

$$= \left(2.2 \times 10^{-4}\right)\left(0.20 \ \text{kg}\right) = 4.4 \times 10^{-5} \ \text{kg}$$

so the wear rate was

$$\text{rate of wear} = \frac{4.4 \times 10^{-5} \ \text{kg}}{1\,000 \ \text{h}} = 4.4 \times 10^{-8} \ \text{kg/h} \qquad \Diamond$$

Chapter 30
Nuclear Energy and Elementary Particles

NOTES FROM SELECTED CHAPTER SECTIONS

30.1 Nuclear Fission

Nuclear fission occurs when a heavy nucleus, such as ^{235}U, splits, or fissions, into two smaller nuclei. In such a reaction, **the total rest mass of the products is less than the original rest mass**. The sequence of events in the fission process is:

- The ^{235}U nucleus captures a thermal (low energy) neutron.

- This capture results in the formation of ^{236}U* (* indicates excited state), and the excess energy of this nucleus causes it to undergo violent oscillations.

- The ^{236}U* nucleus becomes highly elongated, and the force of repulsion between protons in the two halves of the dumbbell-shaped nucleus tends to increase the distortion.

- The nucleus splits into two fragments, emitting several neutrons in the process.

30.2 Nuclear Reactors

A nuclear reactor is a system designed to maintain a **self-sustained chain reaction**. The **reproduction constant** K is defined as the average number of neutrons released from each fission event that will cause another event. In a power reactor, it is necessary to maintain a value of K close to 1. Under this condition, the reactor is said to be **critical**.

Important factors and processes relative to the design and operation of a nuclear reactor include:

- **neutron leakage** — Some neutrons produced by a fission event will escape the volume of the reactor core before producing another fission event. The fraction of neutrons lost by leakage increases as the ratio of the surface area to the volume of the reactor vessel increases.

- **regulating neutron energy** — The fast neutrons produced by fission are reduced in energy by allowing them to undergo collisions with a material of low atomic number (moderator). Slow neutrons have a greater probability than fast neutrons of initiating subsequent fission events.

- **nonfissioning nuclei** — Some neutrons are captured by nuclei that do not undergo fission. The probability that a capture event will occur is reduced when the neutron energies are lower.

- **power level** — The power level of the reactor is controlled by adjusting the number and position of control rods made of materials that are efficient neutron absorbers.

- **reactor safety** — The primary operational concerns related to reactor safety are: containment of the fuel and radioactive fission products, flow of coolant water to the core, disposal of spent fuel rods, and transportation of fuel and waste products.

30.3 Nuclear Fusion

Nuclear fusion is a process in which two light nuclei combine to form a heavier nucleus. The mass of the heavy nucleus is less than the sum of the masses of the two light nuclei; this loss of mass is the source of energy released in a fusion event. The major obstacle in obtaining useful energy from fusion is the large Coulomb repulsive force between charged nuclei as they approach close separation. Sufficient energy must be supplied to the particles to overcome this Coulomb barrier and thereby enable the nuclear attractive force to take over. Important considerations in the effort to achieve controlled fusion are:

- **Ignition temperature** — temperature at which hydrogen nuclei have sufficient kinetic energy to overcome the Coulomb force of repulsion.

- **Ion density** — number of electrons and positive ions per cm^3 which are produced in the gas (plasma) at high temperature.

- **Confinement time** — time that the interacting ions in the plasma are maintained at a temperature high enough for the fusion reaction to proceed.

- **Lawson's criterion** — condition on plasma density and confinement time under which net power output is possible.

30.5 The Fundamental Forces in Nature

There are four fundamental forces in nature:

- The **strong force** is responsible for binding quarks to form neutrons and protons and is the nuclear force that binds neutrons and protons into nuclei. It is a very short-range force and is negligible for separations greater than the approximate size of the nucleus.

- The **electromagnetic force** is responsible for the binding of atoms and molecules. It is a long-range force that decreases in strength as the inverse square of the separation between interacting particles.

- The **weak force** is a short-range nuclear force that tends to produce instability in certain nuclei. The weak interaction governs the structure of basic matter particles and is responsible for beta decay. The electromagnetic and weak forces are now believed to be manifestations of a single force called the **electroweak** force.

- The **gravitational force** is the weakest of all the fundamental forces. It is a long-range force that holds the planets, stars, and galaxies together, and its effect on elementary particles is negligible.

We model the universe as composed of field particles and matter particles. The exchange of field particles mediates the interactions of the matter particles.

- **Gluons** are the field particles for the strong force.

- **Photons** are exchanged by charged particles in electromagnetic interactions.

- **W^+, W^-, and Z bosons** mediate the weak force.

- **Gravitons** are hypothetical quantum particles of the gravitational field.

30.6 Positrons and Other Antiparticles

A particle and its antiparticle (for example an electron and a positron) have the same mass but opposite charge. Other properties of the pair may have opposite values such as lepton number and baryon number.

Pair production is a process in which a particle and its antiparticle are simultaneously produced as a photon disappears. The minimum energy of the photon must be equal to the total rest energy of the created pair. Pair production must occur in the vicinity of a large nucleus in order to conserve momentum.

Pair annihilation is an event in which a particle-antiparticle pair initially at rest, combine with each other, disappear, and produce two photons which move in opposite directions with equal energy and with the same magnitude of momentum.

30.8   Classification of Particles

All particles (other than photons) can be classified into two categories: **hadrons** and **leptons**.

Hadrons are composed of quarks and interact via all fundamental forces. They include two classes of particles grouped according to their masses and spins:

*   **Mesons** — decay finally into electrons, positrons, neutrinos, and photons; they have spin quantum number 0 or 1.

*   **Baryons** — (with the exception of the proton) decay in a manner leading to end products which include a proton; they have spin quantum number $\frac{1}{2}$ or $\frac{3}{2}$.

Leptons (e^-, μ^-, τ^-, ν_e, ν_μ, ν_τ and their antiparticles) interact by the weak interaction, the electromagnetic interaction (if charged), and (presumably) the gravitational interaction but not by the strong force. Leptons have spin number $\frac{1}{2}$. Electrons, muons and neutrinos are included in this group.

30.9   Conservation Laws
30.10 Strange Particles and Strangeness

Conservation laws in the study of elementary particles are based on empirical evidence:

*   **Conservation of baryon number** — Whenever a nuclear reaction or decay occurs, the sum of the baryon numbers before the process must equal the sum of the baryon numbers after the process. This requirement is quantified by assiging a baryon number, B ($B = +1$ for baryons, $B = -1$ for antibaryons, $B = 0$ for all other particles).

*   **Conservation of lepton number** — For each variety there are three conservation laws, quantified by assigning three lepton numbers, one for each variety of lepton: electron-lepton number, L_e; muon-lepton number, L_μ; and tau-lepton number, L_τ. In each case the sum of the lepton numbers before a reaction or decay must equal the sum of the lepton numbers after the reaction or decay.

- **Conservation of strangeness** — Strange particles are always produced in pairs by the strong interaction and decay very slowly, as is characteristic of the weak interaction. Strange particles are assigned a **strangeness quantum number**, S; quantitatively $S = \pm 1, \pm 2, \pm 3$ for strange particles and $S = 0$ for nonstrange particles. Whenever a nuclear reaction or decay occurs via the strong or electromagnetic interactions, the sum of the strangeness numbers before the process must equal the sum of the strangeness numbers after the process ($\Delta S = 0$). Decays occurring via the weak interaction include the loss of one strange particle. This process proceeds slowly and violates the law of conservation of strangness.

30.12 Quarks

Quarks and antiquarks (of six possible types or flavors) make up baryons and mesons; the identity of each particle is determined by the particular combination of quarks. *Baryons consist of three quarks, antibaryons consist of three antiquarks, and mesons consist one quark and one antiquark.* The table at right lists the charge and baryon number for each of the six quarks and antiquarks.

Quark, antiquark	Charge number	Baryon
Up: $u, \bar{u}$	$+\frac{2}{3}e, -\frac{2}{3}e$	$\frac{1}{3}, -\frac{1}{3}$
Down: $d, \bar{d}$	$-\frac{1}{3}e, +\frac{1}{3}e$	$\frac{1}{3}, -\frac{1}{3}$
Strange: $s, \bar{s}$	$-\frac{1}{3}e, +\frac{1}{3}e$	$\frac{1}{3}, -\frac{1}{3}$
Charm: $c, \bar{c}$	$+\frac{2}{3}e, -\frac{2}{3}e$	$\frac{1}{3}, -\frac{1}{3}$
Bottom: $b, \bar{b}$	$-\frac{1}{3}e, +\frac{1}{3}e$	$\frac{1}{3}, -\frac{1}{3}$
Top: $t, \bar{t}$	$+\frac{2}{3}e, -\frac{2}{3}e$	$\frac{1}{3}, -\frac{1}{3}$

The strong force between quarks is referred to as the **color force** and is mediated by massless particles called **gluons**. **Color charge** (red, green , blue) is a property assigned to quarks, which allows combinations of quarks to satisfy the exclusion principle.

EQUATIONS AND CONCEPTS

The **fission of an uranium nucleus** by bombardment with a low energy neutron results in the production of **fission fragments** and typically two or three neutrons. The energy released in the fission event appears in the form of kinetic energy of the fission fragments and the neutrons.

$$\begin{aligned} {}^{1}_{0}n + {}^{235}_{92}U &\rightarrow {}^{236}_{92}U^* \\ &\rightarrow X + Y + \text{neutrons} \end{aligned}$$

(30.1)

U^* is a short-lived intermediate state

Mass and energy conservation requirements are satisfied by many different combinations of fission fragments. On the average, 2.47 neutrons are produced per fission event. Equation 30.2 shows a typical fission reaction.

$$_0^1n + _{92}^{235}U \rightarrow _{56}^{141}Ba + _{36}^{92}Kr + 3\,_0^1n \qquad (30.2)$$

The **fusion reactions** shown here are those most likely to be used as the basis for the design and operation of a fusion power reactor. The Q values refer to the energy released in each reaction.

$$_1^2D + _1^2D \rightarrow _2^3He + _0^1n \qquad (30.4)$$
$$(Q = 3.27 \text{ MeV})$$
$$_1^2D + _1^2D \rightarrow _1^3T + _1^1H$$
$$(Q = 4.03 \text{ MeV})$$
$$_1^2D + _1^3T \rightarrow _2^4He + _0^1n$$
$$(Q = 17.59 \text{ MeV})$$

Lawson's criterion states the conditions under which a net power output of a fusion reactor is possible. In these expressions, n is the plasma density (number of ions per cubic cm), and τ is the plasma confinement time (the time during which the interacting ions are maintained at a temperature equal to or greater than that required for the reaction to proceed).

$$n\tau \geq 10^{14} \text{ s/cm}^3 \qquad (30.5)$$
$$(D - T \text{ interaction})$$

$$n\tau \geq 10^{16} \text{ s/cm}^3$$
$$(D - D \text{ interaction})$$

Three varieties of pions correspond to three charge states: π^+, π^-, and π^0. Pions are very unstable particles. Equation 30.6 shows the decay of π^- (with a lifetime of $\sim 2.6 \times 10^{-8}$ s).

$$\pi^+ \rightarrow \pi^0 + e^+ + \nu_e$$
$$\pi^- \rightarrow \mu^- + \bar{\nu} \qquad (30.6)$$
$$\pi^0 \rightarrow \gamma + \gamma$$

Muons are in two varieties. Example decay processes are shown.

$$\mu^+ \rightarrow e^+ + \nu_e + \bar{\nu}_\mu$$
$$\mu^- \rightarrow e^- + \nu_\mu + \bar{\nu}_e$$

Pion transfer in a proton-neutron interaction, via the strong force, occurs in a time that can be can be obtained from the uncertainty principle, $\Delta t \Delta E \approx \hbar$.

$$\Delta t \approx \frac{\hbar}{\Delta E} = \frac{\hbar}{m_\pi c^2} \qquad (30.7)$$

An order of magnitude of **the range of the weak and strong forces** can be found by assuming that the speed of the particle mediating the exchange equals the speed of light. Equation 30.8 gives the range for a pion exchange between a neutron and a proton.

$$d \approx \frac{\hbar}{m_\pi c} \qquad (30.8)$$

REVIEW CHECKLIST

▷ Describe the sequence of events which occur during the fission process. Write an equation which represents a typical fission event. Use data obtained from the binding energy curve to estimate the disintegration energy of a typical fission event.

▷ List the major parameters influencing fission reactor design and operation which are important in maintaining a steady power level.

▷ Describe the basis of energy release in fusion and write out several nuclear reactions which might be used in a fusion-powered reactor.

▷ Outline the classification of elementary particles and mention several characteristics of each group (relative mass value, spin, decay mode).

▷ Determine whether or not a suggested decay can occur based on the conservation of baryon number and the conservation of lepton number.

▷ Identify particles corresponding to a given quark state.

▷ Show that in a given reaction (e.g. involving pions and protons) the net number of each quark type is conserved.

▷ Use conservation laws to identify reactants and products in a proposed reaction.

▷ Calculate the minimum photon energy required to produce a given particle-antiparticle pair.

SOLUTIONS TO SELECTED END-OF-CHAPTER PROBLEMS

5.   Assume that ordinary soil contains natural uranium in amounts of 1 part per million by mass. (a) How much uranium is in the top 1.00 meter of soil on a 1-acre $\left(43\,560\text{ ft}^2\right)$ plot of ground, assuming the specific gravity of soil is 4.00? (b) How much of the isotope ^{235}U, appropriate for nuclear reactor fuel, is in this soil? [*Hint*: See Appendix B for the percent abundance of $^{235}_{92}\text{U}$.]

Solution

(a)   The volume of soil in the top 1.00 meter on a one acre plot is

$$V = Ah = \left[43\,560\text{ ft}^2 \left(\frac{1\text{ m}}{3.281\text{ ft}} \right)^2 \right] (1.00\text{ m}) = 4.05 \times 10^3\text{ m}^3$$

With a specific gravity of 4.00, this soil is four times as dense as water. Thus, the mass of soil in the top 1.00 m on a one acre plot is

$$m = \rho V = \left(4.00 \rho_{water} \right) V$$

$$= 4.00 \left(1.00 \times 10^3\text{ kg/m}^3 \right) \left(4.05 \times 10^3\text{ m}^3 \right) = 1.62 \times 10^7\text{ kg}$$

Since natural uranium makes up one part per million by mass of ordinary soil, the mass of uranium in this soil is

$$m_{uranium} = \frac{m}{1.00 \times 10^6} = \frac{1.62 \times 10^7\text{ kg}}{1.00 \times 10^6} = 16.2\text{ kg} \qquad \Diamond$$

(b)   The percentage abundance of the isotope ^{235}U in naturally occurring uranium is 0.720%. Thus, the mass of this isotope found in the 16.2 kg of natural uranium from the soil in the upper 1.00 m on a one acre plot is

$$m_{^{235}\text{U}} = \left(\frac{0.720}{100} \right) m_{uranium} = \left(\frac{0.720}{100} \right) (16.2\text{ kg}) = 0.117\text{ kg} = 117\text{ g} \qquad \Diamond$$

9.   An all-electric home uses approximately $2\,000$ kWh of electric energy per month. How much uranium-235 would be required to provide this house with its energy needs for one year? (Assume 100% conversion efficiency and 208 MeV released per fission.)

Solution

The total energy required to meet the energy needs of this house for one year is

$$E = \left(2\,000 \ \frac{\text{kWh}}{\text{month}}\right)\left(3.60 \times 10^{6} \ \frac{\text{J}}{\text{kWh}}\right)\left(12 \ \frac{\text{months}}{\text{yr}}\right) = 8.64 \times 10^{10} \ \text{J/yr}$$

Thus, assuming 100% conversion efficiency and 208 MeV released per fission, the number of fission events required per year is

$$N = \frac{E}{208 \ \text{MeV/event}} = \frac{8.64 \times 10^{10} \ \text{J/yr}}{208 \ \text{MeV/event}} \left(\frac{1 \ \text{MeV}}{1.60 \times 10^{-13} \ \text{J}}\right) = 2.60 \times 10^{21} \ \frac{\text{events}}{\text{yr}}$$

The number of moles of ^{235}U that will be required each year is then

$$n = \frac{N}{\text{Avogadro's number}} = \frac{2.60 \times 10^{21} \ \text{atoms/yr}}{6.02 \times 10^{23} \ \text{atoms/mol}} = 4.31 \times 10^{-3} \ \text{mol/yr}$$

Since 1 mole of ^{235}U has a mass of 235 g, the mass of ^{235}U required to supply the energy needs of this home for one year is

$$m = nM = \left(4.31 \times 10^{-3} \ \text{mol/yr}\right)\left(235 \ \text{g/mol}\right) = 1.01 \ \text{g/yr} \qquad\qquad \Diamond$$

11.   When a star has exhausted its hydrogen fuel, it may fuse other nuclear fuels. At temperatures above 1.0×10^8 K, helium fusion can occur. Write the equations for the following processes: (a) Two alpha particles fuse to produce a nucleus A and a gamma ray. What is nucleus A? (b) Nucleus A absorbs an alpha particle to produce a nucleus B and a gamma ray. What is nucleus B? (c) Find the total energy released in the reactions given in (a) and (b). [*Note*: The mass of $^8_4\text{Be} = 8.005\ 305$ u .]

Solution

(a)   The first reaction in this sequence is $^4_2\text{He} + {}^4_2\text{He} \rightarrow {}^A_Z\text{X} + \gamma$

  where ^A_ZX is the unknown product nucleus A.

  Requiring that charge be conserved gives $2 + 2 = Z + 0$

  or $Z = 4$, meaning that A is an isotope of Beryllium.

  Conserving the number of nucleons gives $4 + 4 = A + 0$

  or $A = 8$. Thus, the product nucleus is ^8_4Be.   ◊

(b)   The next reaction in the sequence is $^8_4\text{Be} + {}^4_2\text{He} \rightarrow {}^A_Z\text{X} + \gamma$

  where ^A_ZX is now the unknown product nucleus B.

  Requiring that charge be conserved gives $4 + 2 = Z + 0$

  or $Z = 6$, meaning that B is an isotope of Carbon.

  Conserving the number of nucleons gives $8 + 4 = A + 0$

  or $A = 12$. Thus, the product nucleus is $^{12}_6\text{C}$.   ◊

(c)   The net result of this pair of reactions is to fuse three alpha particles into a carbon-12 nucleus along with the production of two gamma rays. The overall mass deficit is

$$\Delta m = 3m_{^4_2\text{He}} - m_{^{12}_6\text{C}} = 3\left(4.002\ 602\ \text{u}\right) - 12.000\ 000\ \text{u} = 0.007\ 806\ \text{u}$$

The total energy released in this pair of reactions is then

$$Q = (\Delta m)c^2 = \left(0.007\ 806\ \text{u}\right)\left(931.5\ \text{MeV/u}\right) = 7.27\ \text{MeV}$$ ◊

17.   A photon produces a proton-antiproton pair according to the reaction $\gamma \rightarrow p + \bar{p}$. What is the minimum possible frequency of the photon? What is its wavelength?

Solution

Note that pair production cannot occur in a vacuum. It must occur in the presence of a massive particle which can absorb at least some of the momentum of the photon and allow total linear momentum to be conserved.

When a particle-antiparticle pair is produced by a photon having the minimum possible frequency, and hence minimum possible energy, the nearby massive particle absorbs all the momentum of the photon, allowing both components of the particle-antiparticle pair to be left at rest. In such an event, the total kinetic energy afterwards is essentially zero, and the photon need only supply the total rest energy of the pair produced. The minimum photon energy required to produce a proton-antiproton pair is

$$E_\gamma = 2(E_R)_{proton} = 2(938.3 \text{ MeV})(1.60 \times 10^{-13} \text{ J/MeV}) = 3.00 \times 10^{-10} \text{ J}$$

The frequency of this photon is

$$f = \frac{E_\gamma}{h} = \frac{3.00 \times 10^{-10} \text{ J}}{6.63 \times 10^{-34} \text{ J} \cdot \text{s}} = 4.53 \times 10^{23} \text{ Hz} \qquad \Diamond$$

and its wavelength is

$$\lambda = \frac{c}{f} = \frac{3.00 \times 10^8 \text{ m/s}}{4.53 \times 10^{23} \text{ Hz}} = 6.62 \times 10^{-16} \text{ m} = 0.662 \text{ fm} \qquad \Diamond$$

20. Calculate the order of magnitude of the range of the force that might be produced by the virtual exchange of a proton.

Solution

The virtual exchange of a proton by two particles would violate the principle of conservation of energy by the amount ΔE required to create the proton. This violation of a fundamental conservation law could occur within the limits set by the uncertainty principle $\Delta E \Delta t \geq \hbar/2$ if the proton ceases to exist within a time $\Delta t \approx \hbar/\Delta E$.

Thus, we may approximate the upper limit on the lifetime of a virtual proton. The minimum uncertainty in the energy is

$$\Delta E = m_p c^2 = 938.3 \text{ MeV}$$

so, the maximum time the virtual proton could exist and violate conservation of energy would be

$$\Delta t \approx \frac{\hbar}{\Delta E} = \frac{1.05 \times 10^{-34} \text{ J} \cdot \text{s}}{938.3 \text{ MeV}} \left(\frac{1 \text{ MeV}}{1.60 \times 10^{-13} \text{ J}} \right) = 7.0 \times 10^{-25} \text{ s}$$

The maximum distance a virtual proton could travel in its lifetime, and hence an estimate for the range of the force that might be produced by the virtual exchange of a proton, would be

$$d \approx c(\Delta t) = (3.00 \times 10^8 \text{ m/s})(7.0 \times 10^{-25} \text{ s}) = 2.1 \times 10^{-16} \text{ m}$$

or, the order of magnitude for the range of this force would be

$$d \sim 10^{-16} \text{ m}$$

◊

25. Identify the unknown particle on the left side of the reaction

$$? + p \rightarrow n + \mu^+$$

Solution

The proton, neutron and μ^+ are hadrons (that is, particles which interact via the strong interaction). The strong interaction always conserves several quantities, namely charge, Q; baryon number, B; electron-lepton number, L_e; muon-lepton number, L_μ; tau-lepton number, L_τ; and strangeness, S. These conservation laws may be used to identify the unknown particle, X, in the reaction $X + p \rightarrow n + \mu^+$. The table below summarizes the properties of the particles present before and after the reaction (see Table 30.2 in the textbook). Note that the property values of an antiparticle are the negative of those for the corresponding particle.

	Particles before reaction			Particles after reaction		
Property	Particle X	Proton	Total	Neutron	Antimuon	Total
Q	Q_X	$+e$	$Q_X + e$	0	$+e$	$+e$
B	B_X	$+1$	$B_X + 1$	$+1$	0	$+1$
L_e	L_{eX}	0	L_{eX}	0	0	0
L_μ	$L_{\mu X}$	0	$L_{\mu X}$	0	-1	-1
L_τ	$L_{\tau X}$	0	$L_{\tau X}$	0	0	0
S	S_X	0	S_X	0	0	0

If the reaction is to conserve each of the properties listed, the entries in the total column before reaction must equal the entries in the total column after reaction. Thus, $Q_X + e = +e$, $B_X + 1 = +1$, $L_{eX} = 0$, $L_{\mu X} = -1$, $L_{\tau X} = 0$, and $S_X = 0$. The property values for the unknown particle are therefore,

$$Q_X = 0, \ B_X = 0, L_{eX} = 0, L_{\mu X} = -1, L_{\tau X} = 0, \text{ and } S_X = 0$$

The particle with these properties is $\bar{\nu}_\mu$, the antimuon-neutrino. ◊

30.   A K^0 particle at rest decays into a π^+ and a π^-. What will be the speed of each of the pions? The mass of the K^0 is $497.7 \text{ MeV}/c^2$, and the mass of each pion is $139.6 \text{ MeV}/c^2$.

Solution

With the K^0 at rest before the decay, both the linear momentum and the kinetic energy before the decay are zero. Conservation of linear momentum requires that the total linear momentum after the decay be zero. This means that the two pions present after the decay must travel in opposite directions with equal magnitude momenta. That is, $p_{\pi^+} = p_{\pi^-}$.

Since the two pions have equal masses, they also have equal rest energies, $E_R = mc^2$. Then, from $E^2 = p^2c^2 + E_R^2$, we see that the pions will have equal total energies as well as equal kinetic energies, $KE = E - E_R$.

The rest energy converted into kinetic energy in the decay is

$$Q = (\Delta m)c^2 = \left(m_{K^0} - m_{\pi^+} - m_{\pi^-}\right)c^2 = 497.7 \text{ MeV} - 2(139.6 \text{ MeV}) = 218.5 \text{ MeV}$$

Thus, the kinetic energy of each pion after the decay is

$$KE = KE_{\pi^+} = KE_{\pi^-} = \frac{KE_{before} + Q}{2} = \frac{0 + 218.5 \text{ MeV}}{2} = 109.3 \text{ MeV}$$

and the total energy of each pion is

$$E = E_R + KE = 139.6 \text{ MeV} + 109.3 \text{ MeV} = 248.9 \text{ MeV}$$

The speed of either pion can then be found from $\qquad E = \dfrac{E_R}{\sqrt{1 - (v/c)^2}}$

This gives $v = c\sqrt{1 - \left(\dfrac{E_R}{E}\right)^2} = c\sqrt{1 - \left(\dfrac{139.6 \text{ MeV}}{248.9 \text{ MeV}}\right)^2} = 0.827\,8\,c$ ◊

35. Find the number of electrons, and of each species of quark in 1 L of water.

Solution

An ordinary water molecule (H_2O) contains two neutral atoms of ^1_1H and one neutral atom of $^{16}_8O$. Thus, each molecule contains 10 protons, 10 electrons, and 8 neutrons.

The molecular weight of water is 18.0 g/mol and its density is

$$\rho = 1.00 \times 10^3 \ \frac{kg}{m^3} \left(\frac{10^3 \ g}{1 \ kg}\right)\left(\frac{1 \ m^3}{10^3 \ L}\right) = 1.00 \times 10^3 \ g/L$$

Therefore, the mass of one liter of water is $1.00 \times 10^3 \ g$, and the number of molecules it contains is

$$N = \left(\frac{m}{M}\right) N_A = \left(\frac{1.00 \times 10^3 \ g}{18.0 \ g/mol}\right)\left(6.02 \times 10^{23} \ \frac{molecules}{mol}\right) = 3.34 \times 10^{25} \ molecules$$

The number of electrons, protons, and neutrons in the one liter of water is then

$$N_e = \left(10 \ \frac{electrons}{molecule}\right)(3.34 \times 10^{25} \ molecules) = 3.34 \times 10^{26} \ electrons \qquad ◊$$

$$N_p = \left(10 \ \frac{protons}{molecule}\right)(3.34 \times 10^{25} \ molecules) = 3.34 \times 10^{26} \ protons$$

and

$$N_n = \left(8 \ \frac{neutrons}{molecule}\right)(3.34 \times 10^{25} \ molecules) = 2.68 \times 10^{26} \ neutrons$$

Each proton (uud) contains 2 up quarks and 1 down quark, while each neutron (udd) has 1 up quark and 2 down quarks. Thus, the one liter of water will contain

$$N_u = 2N_p + N_n = 2(3.34 \times 10^{26}) + 2.68 \times 10^{26} = 9.36 \times 10^{26} \ up \ quarks \qquad ◊$$

and

$$N_d = N_p + 2N_n = 3.34 \times 10^{26} + 2(2.68 \times 10^{26}) = 8.70 \times 10^{26} \ down \ quarks \qquad ◊$$

41. A Σ^0 particle traveling through matter strikes a proton, and a Σ^+, and a gamma ray, as well as a third particle, emerges. Use the quark model of each to determine the identity of the third particle.

Solution

The reaction of interest is $\quad \Sigma^0 + p \rightarrow \Sigma^+ + \gamma + X$

where X is an unknown particle. We recognize that the Σ^0 particle and the proton are both baryons and interact via the strong interaction. Thus, we could attempt to identify the unknown particle by using the fact that the strong interaction always obeys several conservations laws: conservation of charge, conservation of baryon number, conservation of strangeness, and conservation of lepton numbers (one for each variety of lepton). However, we shall consider the quark composition (in terms of up, down, and strange quarks) of the known particles before and after the reaction to find the identity of particle X.

Consider Table 30.4 in the textbook to see that the quark composition of a Σ^0 particle is uds, that of a proton is uud, and that of a Σ^+ particle is uus. Thus, in terms of the quark composition, the reaction is

$$\text{uds} + \text{uud} \rightarrow \text{uus} + (0 \text{ quarks}) + (n_u n_d n_s)$$

where n_u, n_d, and n_s represents the number of up, down, and strange quarks respectively, contained in the mystery particle X. Note that before the reaction, there is a total of 3 up quarks, 2 down quarks, and 1 strange quark present. After the reaction, we have: $n_u + 2$ up quarks, $n_d + 0$ down quarks, and $n_s + 1$ strange quarks present.

In order to conserve the net number of each type of quark in this reaction, we must have:

$$n_u + 2 = 3 \quad \text{or} \quad n_u = 1 \qquad \text{so } X \text{ must have 1 up quark}$$

$$n_d + 0 = 2 \quad \text{or} \quad n_d = 2 \qquad \text{so } X \text{ must have 2 down quarks}$$

and $\qquad n_s + 1 = 1 \quad \text{or} \quad n_s = 0 \qquad \text{so } X \text{ must have 0 strange quarks}$

Thus, the quark composition of particle X must be udd and from Table 30.4 we see that particle X must be a neutron. ◊

45.   Find the energy released in the fusion reaction

$$_1^1\text{H} + _2^3\text{He} \rightarrow _2^4\text{He} + e^+ + \nu_e$$

Solution

If we had nuclear masses (that is, masses of atoms stripped bare of all electrons), the equation for the energy released in the fusion reaction would be

$$Q = (\Delta m)c^2 = \left(m_{_1^1\text{H}} + m_{_2^3\text{He}} - m_{_2^4\text{He}} - m_e\right)c^2$$

where we have assumed that the neutrino has zero mass.

However, the masses given in Appendix B are those of neutral atoms. Thus, we must be very careful how we handle electron masses, especially in light of the fact that one of the product particles (the positron) in this reaction has the same mass as an electron.

To obtain neutral atoms, add three electrons to both sides of the reaction equation. We then have

$$_1^1\text{H}_{\text{atom}} + _2^3\text{He}_{\text{atom}} \rightarrow _2^4\text{He}_{\text{atom}} + e^- + e^+ + \nu_e$$

and the equation for the energy released becomes

$$Q = (\Delta m)c^2 = \left(m_{_1^1\text{H}_{\text{atom}}} + m_{_2^3\text{He}_{\text{atom}}} - m_{_2^4\text{He}_{\text{atom}}} - 2m_e\right)c^2$$

Using the masses of neutral atoms from Appendix B and the mass of the electron from Table 29.4 gives the energy released in the fusion reaction as

$$Q = \left[1.007\,825\ \text{u} + 3.016\,029\ \text{u} - 4.002\,602\ \text{u} - 2(0.000\,549\ \text{u})\right]\left(931.5\ \frac{\text{MeV}}{\text{u}}\right)$$

$$Q = 18.8\ \text{MeV} \qquad\qquad\qquad \Diamond$$

49.   The atomic bomb dropped on Hiroshima on August 6, 1945 released 5×10^{13} J of energy (equivalent to that from 12 000 tons of TNT). Estimate (a) the number of $^{235}_{92}U$ nuclei fissioned and (b) the mass of this $^{235}_{92}U$.

Solution

(a)   When a $^{235}_{92}U$ undergoes fission, many different sets of fission products may be produced. Each different type of fission reaction will yield a different amount of energy. We shall assume that, on average, a typical fission event of $^{235}_{92}U$ releases about 200 MeV of energy. Then, the number of fission events required to produce a total of 5×10^{13} J would be

$$N \approx \frac{5 \times 10^{13} \text{ J}}{200 \text{ MeV/event}} \left(\frac{1 \text{ MeV}}{1.60 \times 10^{-13} \text{ J}} \right) = 1.56 \times 10^{24} \text{ events}$$

or, to 1 significant figure, $N \approx 2 \times 10^{24}$ events   ◊

(b)   Each mole of ^{235}U will have a mass of $M = 235$ g and will contain Avogadro's number of fissionable nuclei. From this, the mass of 1.56×10^{24} ^{235}U atoms is found to be

$$m = nM \approx \left(\frac{1.56 \times 10^{24} \text{ atoms}}{6.02 \times 10^{23} \text{ atoms/mol}} \right) (235 \text{ g/mol}) = 6 \times 10^2 \text{ g} = 0.6 \text{ kg}$$ ◊

53.   (a) Show that about 1.0×10^{10} J would be released by the fusion of the deuterons in 1.0 gal of water. Note that 1 out of every 6 500 hydrogen atoms is a deuteron. (b) The average energy consumption rate of a person living in the United States is about 1.0×10^4 J/s (an average power of 10 kW). At this rate, how long would the energy needs of one person be supplied by the fusion of the deuterons in 1.0 gal of water? Assume that the energy released per deuteron is 1.64 MeV.

Solution

(a)   The mass of 1.0 gal of water is

$$m = \rho V = \left(1.00 \times 10^3 \ \frac{\text{kg}}{\text{m}^3}\right) \left[1.0 \ \text{gal} \left(\frac{3.786 \ \text{L}}{1 \ \text{gal}}\right) \left(\frac{10^{-3} \ \text{m}^3}{1 \ \text{L}}\right)\right] = 3.8 \ \text{kg}$$

The number of molecules in this quantity of water is

$$N = n N_A = \left(\frac{m}{M}\right) N_A = \left(\frac{3.8 \times 10^3 \ \text{g}}{18 \ \text{g/mol}}\right) \left(6.02 \times 10^{23} \ \text{molecules/mol}\right)$$

or $N = 1.27 \times 10^{26}$ molecules

Each of these water molecules contains two hydrogen nuclei, and one in every 6 500 of these is a deuteron. Thus, the number of deuterium nuclei $\left(^2_1\text{H}\right)$ in the gallon of water is

$$N_\text{d} = \frac{2N}{6 \ 500} = \frac{2\left(1.27 \times 10^{26}\right)}{6 \ 500} = 3.9 \times 10^{22} \ \text{deuterons}$$

Assuming that the energy released per deuteron is 1.64 MeV, the total energy available from the 1.0 gal of water is

$$E = \left(3.9 \times 10^{22} \ \text{deuterons}\right) \left(1.64 \ \frac{\text{MeV}}{\text{deuteron}}\right) \left(\frac{1.6 \times 10^{-13} \ \text{J}}{1 \ \text{MeV}}\right) = 1.0 \times 10^{10} \ \text{J} \qquad \Diamond$$

(b)   The time this could supply the energy needs of one person is

$$t = \frac{E}{\text{consumption rate}} = \frac{1.0 \times 10^{10} \ \text{J}}{1.0 \times 10^4 \ \text{J/s}} \left(\frac{1 \ \text{d}}{86 \ 400 \ \text{s}}\right) = 12 \ \text{d} \qquad \Diamond$$